Regularity of Free Boundaries in Obstacle-Type Problems

Regularity of Free Boundaries in Obstacle-Type Problems

Arshak Petrosyan
Henrik Shahgholian
Nina Uraltseva

Graduate Studies in Mathematics
Volume 136

American Mathematical Society
Providence, Rhode Island

2010 *Mathematics Subject Classification.* Primary 35R35.

For additional information and updates on this book, visit
www.ams.org/bookpages/gsm-136

Library of Congress Cataloging-in-Publication Data

Petrosyan, Arshak, 1975–
Regularity of free boundaries in obstacle-type problems / Arshak Petrosyan, Henrik Shahgholian, Nina Uraltseva.
p. cm. — (Graduate studies in mathematics ; v. 136)
Includes bibliographical references and index.
ISBN 978-0-8218-8794-3 (alk. paper)
1. Boundaqry value problems. I. Shahgholian, Henrik, 1960– II. Ural′tseva, N. N. (Nina Nikolaevna) III. Title.

QA379.P486 2012
515′.353—dc23

2012010200

∞ The paper used in this book is acid-free and falls within the guidelines established to ensure permanence and durability.
Visit the AMS home page at `http://www.ams.org/`

10 9 8 7 6 5 4 3 2 1 17 16 15 14 13 12

Contents

Preface

Free boundary problems (FBPs) are considered today as one of the most important directions in the mainstream of the analysis of partial differential equations (PDEs), with an abundance of applications in various sciences and real world problems. In the past two decades, various new ideas, techniques, and methods have been developed, and new important, challenging problems in physics, industry, finance, biology, and other areas have arisen.

The study of free boundaries is an extremely broad topic not only due to the diversity of applications but also because of the variety of the questions one may be interested in, ranging from modeling and numerics to the purely theoretical questions. This breadth presents challenges and opportunities!

A particular direction in free boundary problems has been the study of the regularity properties of the solutions and those of the free boundaries. Such questions are usually considered very hard, as the free boundary is not known a priori (it is part of the problem!) so the classical techniques in elliptic/parabolic PDEs do not apply. In many cases the success is achieved by combining the ideas from PDEs with the ones from geometric measure theory, the calculus of variations, harmonic analysis, etc.

Today there are several excellent books on free boundaries, treating various issues and questions: e.g. [**DL76**], [**KS80**], [**Cra84**], [**Rod87**], [**Fri88**], [**CS05**]. These books are great assets for anyone who wants to learn FBPs and related techniques; however, with the exception of [**CS05**], they date back two decades. We believe that there is an urge for a book where some of the most recent developments and new methods in the regularity of free boundaries can be introduced to the nonexperts and particularly to the graduate students starting their research in the field. This gap in the literature has been partially filled by the aforementioned book of Caffarelli and

Salsa [**CS05**], which treats the Stefan-type free boundary problems (with the Bernoulli gradient condition). Part 3 in [**CS05**], in particular, covers several technical tools that should be known to anyone working in the field of PDEs/FBPs.

Our intention, in this book, was to give a coherent presentation of the study of the regularity properties of the free boundary for a particular type of problems, known as *obstacle-type problems*. The book grew out of the lecture notes for the courses and mini-courses given by the authors at various locations, and hence we believe that the format of the book is most suitable for a graduate course (see the end of the Introduction for suggestions). Notwithstanding this, we have to warn the reader that this book is far from being a complete reference for the regularity theory. We hope that it gives a reasonably good introduction to techniques developed in the past two decades, including those due to the authors and their collaborators.

We thank many colleagues and fellow mathematicians for reading parts of this book and commenting, particularly Mark Allen, Darya Apushkinskaya, Mahmoudreza Bazarganzadeh, Paul Feehan, Nestor Guillen, Erik Lindgren, Norayr Matevosyan, Andreas Minne, Sadna Sajadini, Wenhui Shi, Martin Strömqvist. Our special thanks go to Luis Caffarelli, Craig Evans, Avner Friedman, and Juan-Luis Vázquez for their useful suggestions and advice regarding the book.

We gratefully acknowledge the support from the following funding agencies: A.P. was supported in part by the National Science Foundation Grant DMS-0701015; H.S. was supported by the Swedish Research Council; N.U. was supported by RFBR Grant No. 11-01-00825-a NSh.4210.2010.1, and Russian Federal Target Program 2010-1.1-111-128-033.

Finally, we thank the Mathematical Sciences Research Institute (MSRI), Berkeley, CA, for hosting a program on *Free Boundary Problems, Theory and Applications* in Spring 2011, where all three of us were at residence and had a fabulous working environment to complete the book.

West Lafayette, IN, USA—Stockholm, Sweden—St. Petersburg, Russia

February 2012

Introduction

In this book we treat free boundary problems of the type

$$\Delta u = f(x, u, \nabla u) \quad \text{in } D \subset \mathbb{R}^n,$$

where the right-hand side f exhibits a jump discontinuity in its second and/or third variables. The discontinuity set is a priori unknown and therefore is said to be *free*. The prototypical example is the so-called *classical obstacle problem*, which minimizes the energy of a stretched membrane over a given obstacle (see more in §1.1) and in its simplest form can be reformulated as

$$\Delta u = \chi_{\{u>0\}}, \quad u \geq 0, \quad \text{in } D. \tag{0.1}$$

In this case $f(x, u, \nabla u) = \chi_{\{u>0\}}$ is the Heaviside function of u. The *free boundary* here is $\Gamma = \partial\{u > 0\} \cap D$. Also note that the *sign condition* $u \geq 0$ in (0.1) appears naturally in this problem. The classical obstacle problem and its variations have been the subject of intense studies in the past few decades. Today, there is a more or less complete and comprehensive theory for this problem, both from theoretical and numerical points of view.

The motivation for studying free boundary problems in general, and obstacle-type problems in particular, has roots in many applications. Classical applications of these problems originate (predominantly) in engineering sciences, where many problems (sometimes after a major simplification) could be formulated as variational inequalities or more general free boundary problems. In many cases variational inequalities can be viewed as obstacle-type problems, with an additional sign condition, as in the case of the classical obstacle problem. This particular feature significantly simplifies the problem, and most methods, up to the early 1990s, relied heavily on this strong property of solutions.

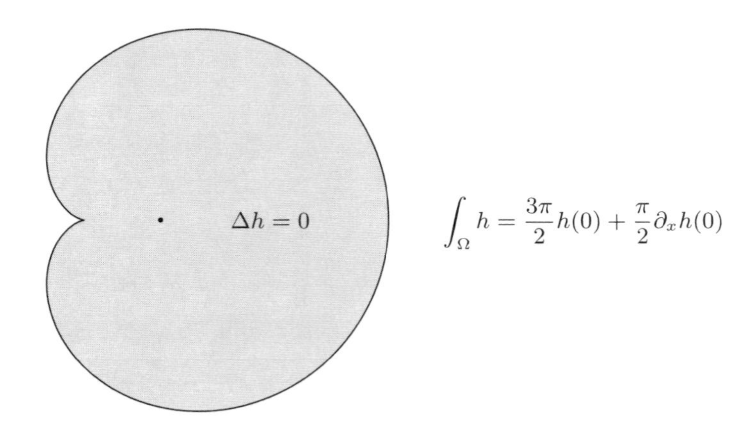

Figure 0.1. Cardioid $\Omega = \{z = w + (1/2)w^2 : |w| < 1\}$ is a quadrature domain

Obstacle-type problems also arise in applications in potential theory, complex analysis, and geophysics. In contrast to variational inequalities, these problems typically lack the sign condition. To manifest, in the simplest form, one such application, we start with the mean-value property for harmonic functions in the unit ball B_1:

$$\int_{B_1} h(y)dy = c_n h(0),$$

where h is any harmonic function, integrable over B_1. More generally, given a distribution μ with compact support (say, a finite set), let Ω be a bounded domain containing $\operatorname{supp}\mu$ such that a generalized mean-value property (or a quadrature formula) holds:

$$\int_\Omega h(y)dy = \int h d\mu,$$

for any integrable harmonic function h in Ω. Such Ω are known as *quadrature domains*. There is an abundance of such domains; see [**Dav74, Sak82**]. An example of a quadrature domain is the so-called cardioid, which can be represented via the conformal mapping $z = w + (1/2)w^2$ of the unit circle (see Fig. 0.1). The distribution μ in this case equals $(\pi/2)(3\delta_0 + \partial_x\delta_0)$, where δ_0 is the Dirac delta function centered at the origin.

By letting $h(y) = |x-y|^{2-n}$, $x \in \Omega^c$, in dimensions $n \geq 3$, and the corresponding logarithmic potential in dimension $n = 2$, one will obtain

$$\int_\Omega |x-y|^{2-n} dy = \int |x-y|^{2-n} d\mu(y), \quad x \in \Omega^c.$$

Now letting

$$u(x) := c_n \int_\Omega |x-y|^{2-n} dy - c_n \int |x-y|^{2-n} d\mu(y),$$

for an appropriately chosen constant c_n, we see that u satisfies

$$\Delta u = \chi_\Omega - \mu, \quad u = |\nabla u| = 0 \text{ in } \Omega^c \quad \text{in } \mathbb{R}^n.$$

In particular, in a small neighborhood $B_r(x^0)$ of any point $x^0 \in \partial\Omega$ we have

$$(0.2) \qquad \Delta u = \chi_\Omega \quad \text{in } B_r(x^0), \quad u = |\nabla u| = 0 \quad \text{in } B_r(x^0) \setminus \Omega.$$

This problem bears resemblance to the classical obstacle problem; however, the important difference is that there is no sign condition imposed on u. More specifically, the solutions of (0.1) are precisely the nonnegative solutions of (0.2).

The boundary of Ω, $\Gamma = \partial\Omega$, is the free boundary in the problem, as we do not know beforehand the location, shape and regularity properties of Γ. Taking Ω to be the cardioid (see Fig. 0.1), we see that Γ may naturally exhibit cusp singularities, but otherwise is smooth (real analytic). It turns out that this is a typical regularity property of free boundaries for obstacle-type problems that we treat in this book.

The main approach to the study of local properties of the free boundary is the so-called method of *blowup*. This method originated in the work of Caffarelli [**Caf80**], motivated by a similar method in geometric measure theory, in the study of minimal surfaces. We will demonstrate this method by the example of the solution u of (0.2). For any $\lambda > 0$ consider the rescalings

$$u_\lambda(x) = u_{x^0,\lambda}(x) = \frac{u(x^0 + \lambda x)}{\lambda^2},$$

which will be defined now in $B_{r/\lambda}$. The factor of λ^2 is chosen so that u_λ still satisfy conditions similar to (0.2):

$$\Delta u_\lambda = \chi_{\Omega_\lambda} \quad \text{in } B_{r/\lambda}, \quad u_\lambda = |\nabla u_\lambda| = 0 \quad \text{in } B_{r/\lambda} \setminus \Omega_\lambda,$$

where

$$\Omega_\lambda = \{x : x^0 + \lambda x \in \Omega\} = \frac{1}{\lambda}(\Omega - x^0).$$

Heuristically, this corresponds to "zooming" with factor $1/\lambda$ near x^0. The idea is now to study the limit as $\lambda \to 0+$ (which would correspond to the idea of "infinite zoom" at x^0). To this end, we need to have some uniform estimates for the family of rescalings $\{u_\lambda\}$. It turns out, in fact, that one can prove the *quadratic growth* of the original solution u near x^0, i.e.,

$$(0.3) \qquad |u(x)| \leq M|x - x^0|^2, \quad x \in B_{r/2}(x^0),$$

which readily implies the uniform estimates

$$|u_\lambda(x)| \leq M|x|^2, \quad x \in B_{r/(2\lambda)},$$

for any $\lambda > 0$, and then the a priori $C^{1,\alpha}$ estimates imply the convergence of u_λ to a certain u_0, locally in the entire space $\mathbb{R}^n$, over a subsequence

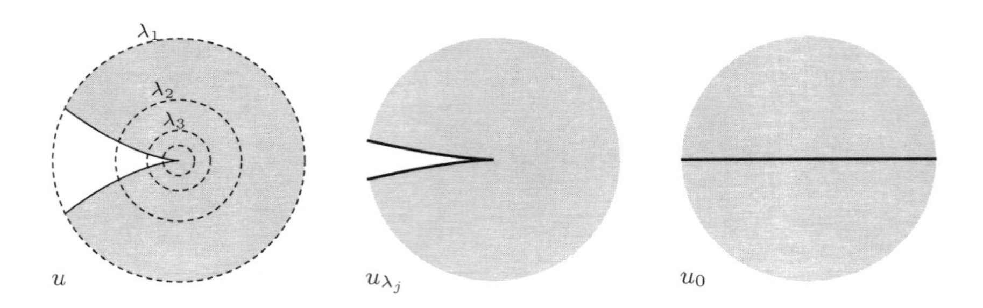

Figure 0.2. Blowup near the cusp point of the cardioid

$\lambda = \lambda_j \to 0+$ (see Fig. 0.2). Such u_0 will be called a *blowup* of u at x^0. There are a few difficulties that one encounters in this process:

1) The quadratic growth estimate is far from being obvious. In fact, if one treats u as being a solution of the equation $\Delta u = \chi_\Omega$ with a bounded right-hand side, then the best estimate one could hope for is

$$|u(x)| \leq M|x - x^0|^2 \log(1/|x - x^0|),$$

and this estimate does not scale well to provide uniform estimates for u_λ. However, using the additional structure of the problem in (0.2), it is possible to remove the logarithmic term. This is relatively easy in the presence of the sign condition $u \geq 0$. However, it requires the application of a powerful monotonicity formula of Alt-Caffarelli-Friedman in the general case (for details see Chapter 2).

2) In principle, the limits over different subsequences $\lambda = \lambda_j \to 0+$ may lead to different blowups. One of the main difficulties is to show that in fact the blowup u_0 is unique.

It is relatively easy to show that the blowups u_0 themselves satisfy a condition similar to (0.2), but in the entire space:

$$\Delta u_0 = \chi_{\Omega_0} \quad \text{in } \mathbb{R}^n, \quad u_0 = |\nabla u_0| = 0 \quad \text{in } \mathbb{R}^n \setminus \Omega_0,$$

or we say that u_0 is a *global solution*. Similarly to the elliptic problems, there is a Liouville-type theorem for global solutions as above, which says for instance that u_0 must either be a quadratic polynomial or must be nonnegative and convex. Different types of blowups lead to a natural classification of free boundary points.

The next step in the study is to transfer some of the properties of the blowups to the rescalings and then to the original solution. For instance, if the blowup u_0 is convex and nonpolynomial, then there is a cone of directions $\mathcal{C}$ such that

$$\partial_e u_0 \geq 0 \quad \text{in } \mathbb{R}^n, \quad e \in \mathcal{C}.$$

It turns out that a slightly modified version of this property can be concluded also for the rescalings u_λ with small $\lambda > 0$, and in particular that for a small $\rho > 0$,

$$\partial_e u \geq 0 \quad \text{in } B_\rho(x^0), \quad e \in \mathcal{C}.$$

The monotonicity in a cone of directions then implies the Lipschitz regularity of Γ. After this initial regularity, one can push the regularity of the free boundary to $C^{1,\alpha}$ by an application of the so-called boundary Harnack principle and then further to $C^{k,\alpha}$ and real analyticity by the so-called method of partial hodograph-Legendre transform. A rigorous, yet readable, implementation of the steps above is the main purpose of this book.

Suggestions for reading/teaching

Anyone who intends to read this book, should have a basic knowledge in elliptic PDEs (maximum/comparison principle, the Harnack inequality, interior/boundary estimates, compactness arguments, etc). We also encourage the reader to look at the book by Caffarelli-Salsa [**CS02**], especially Part 3, for many technical tools that are extensively used in free boundary problems.

The book treats three different but methodologically close problems, here called Problems **A**, **B**, and **C**, and in addition, the so-called thin obstacle problem (Problem **S**). Problem **A** is essentially the same as problem (0.2) described above. For a beginner, we would suggest a study of this problem only, at least in the first reading.

Anyone who wants to give a course in the topic can choose Problem **A**, throughout Chapters 1–6, for a full-time 1/4 semester course. It can also be run over a semester with 2h/week lecture. An extended course can include model Problems **B** and **C** as well. Technically they are not very different, but sometime require finer analysis and new ideas to achieve results similar to those for Problem **A**.

Chapters 7–8 treat singular sets and the touch with fixed boundaries, and can be used as part of an examination where students can make presentations of various parts of these two chapters and do the exercises in these chapters.

In Chapter 9, we treat the so-called *thin obstacle problem* (Problem **S**). This chapter can be taught as a separate course, with some use of existing articles in the field. It can also be used as presentation material for students taking the above format of a course.

Each chapter closes with bibliographical notes for the results in that chapter and some of their generalizations.

At the very end of each chapter, we have tried to gather as many exercises as possible. Proofs of some of the theorems/lemmas are also put in exercises; in many cases hints (or even brief solutions) are provided.

Chapter 1

Model problems

Free boundary problems of obstacle type appear naturally in numerous applications, and the purpose of this chapter is to list some of the most interesting ones, from our point of view (§1.1). Not all of these problems will be treated in detail in this book, but the methods developed here will be applicable to all of them at least with a certain degree of success. For the detailed treatment we have selected three model problems (which we call **A**, **B**, and **C**) that can be put into a more general framework of obstacle-type problems $\mathbf{OT}_1$–$\mathbf{OT}_2$ (§1.2). At the end of this chapter (§1.3) we discuss the almost optimal $W^{2,p}_{\mathrm{loc}} \cap C^{1,\alpha}_{\mathrm{loc}}$ regularity of solutions for any $1 < p < \infty$, $0 < \alpha < 1$.

1.1. Catalog of problems

1.1.1. The classical obstacle problem.

1.1.1.1. *The Dirichlet principle.* A well-known variational principle of Dirichlet says that the solution of the boundary value problem

$$\Delta u = 0 \quad \text{in } D, \qquad u = g \quad \text{on } \partial D,$$

can be found as the minimizer of the (Dirichlet) functional

$$J_0(u) = \int_D |\nabla u|^2 dx,$$

among all u such that $u = g$ on ∂D. More precisely (and slightly more generally), if D is a bounded open set in $\mathbb{R}^n$, $g \in W^{1,2}(D)$ and $f \in L^\infty(D)$, then the minimizer of

$$J(u) = \int_D (|\nabla u|^2 + 2fu)dx \tag{1.1}$$

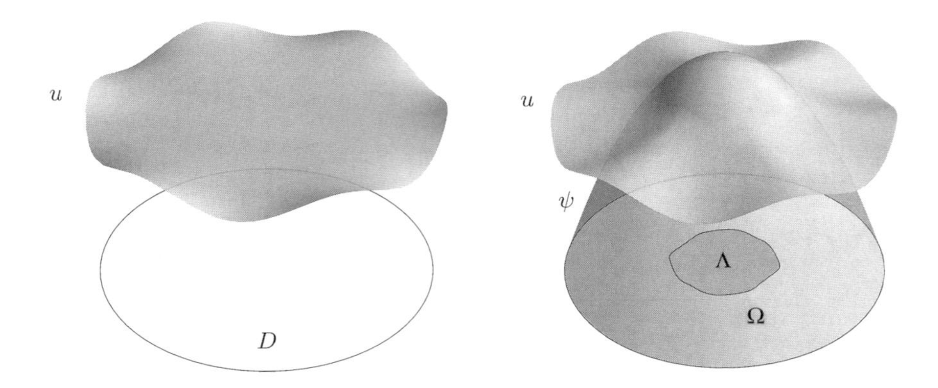

Figure 1.1. Free membrane and the solution of the obstacle problem

over the set

$$\mathfrak{K}_g = \{u \in W^{1,2}(D) : u - g \in W_0^{1,2}(D)\}$$

solves the Poisson equation

$$-\Delta u + f = 0 \quad \text{in } D, \qquad u = g \quad \text{on } \partial D,$$

in the sense of distributions, i.e.

$$\int_D (\nabla u \nabla \eta + f\eta)dx = 0,$$

for all test functions $\eta \in C_0^\infty(D)$ (and more generally for all $\eta \in W_0^{1,2}(D)$). One can think of the graph of u as a membrane attached to a thin wire (the graph of g over ∂D).

1.1.1.2. *The classical obstacle problem.* Suppose now that we are given a certain function $\psi \in C^2(D)$, known as the *obstacle*, satisfying the compatibility condition $\psi \leq g$ on ∂D in the sense that $(\psi - g)_+ \in W_0^{1,2}(D)$. Consider then the problem of minimizing the functional (1.1), but now over the constrained set

$$\mathfrak{K}_{g,\psi} = \{u \in W^{1,2}(D) : u - g \in W_0^{1,2}(D), u \geq \psi \text{ a.e. in } D\}.$$

Since J is continuous and strictly convex on a convex subset $\mathfrak{K}_{g,\psi}$ of the Hilbert space $W^{1,2}(D)$, it has a unique minimizer on $\mathfrak{K}_{g,\psi}$.

As before, we may think of the graph of u as a membrane attached to a fixed wire, which is now forced to stay above the graph of ψ. A new feature in this problem is that the membrane can actually touch the obstacle; i.e. the set

$$\Lambda = \{u = \psi\},$$

known as the *coincidence set*, may be nonempty (see Fig. 1.1). We also denote

$$\Omega = D \setminus \Lambda.$$

The boundary

$$\Gamma = \partial\Lambda \cap D = \partial\Omega \cap D$$

is called the *free boundary*, since it is not known a priori. The study of the free boundary in this and related problems is the main objective in this book.

To obtain the conditions satisfied by the minimizer u, we observe that using the method of regularization, which we discuss in §1.3.2 (or, alternatively, the method of penalization; see Exercise 1.10 or Friedman [**Fri88**, Chapter 1]) one can show that the minimizer is not only in $W^{1,2}(D)$, but actually is in $W^{2,p}_{\mathrm{loc}}(D)$ for any $1 < p < \infty$ and consequently (by the Sobolev embedding theorem) is in $C^{1,\alpha}_{\mathrm{loc}}(D)$ for any $0 < \alpha < 1$. Then it is straightforward to show (see Exercise 1.1) that

$$\begin{aligned} \Delta u &= f && \text{in } \Omega = \{u > \psi\}, \\ \Delta u &= \Delta\psi && \text{a.e. on } \Lambda = \{u = \psi\}. \end{aligned}$$

Besides,

$$\Delta u \le f \quad \text{in } D$$

in the sense of distributions. Combining the properties above, we obtain that the solution of the obstacle problem is a function $u \in W^{2,p}_{\mathrm{loc}}(D)$ for any $1 < p < \infty$, which satisfies

$$-\Delta u + f \ge 0, \quad u \ge \psi, \quad (-\Delta u + f)(u - \psi) = 0 \quad \text{a.e. in } D, \tag{1.2}$$

$$u - g \in W^{1,2}_0(D). \tag{1.3}$$

This is known as the *complementarity problem* and uniquely characterizes the minimizers of J over $\mathfrak{K}_{g,\psi}$. Complementarity condition (1.2) is also written quite often as

$$\min\{-\Delta u + f, u - \psi\} = 0.$$

1.1.1.3. *Reduction to the case of zero obstacle.* Since the governing operator (Δ) is linear, it is possible to reduce the problem to the case when the obstacle is identically 0. Indeed, for a solution u of the obstacle problem, consider the difference $v = u - \psi$. Then it is straightforward to see that v is the minimizer of the functional

$$J_1(v) = \int_D (|\nabla v|^2 + 2f_1 v)dx$$

over the set $\mathfrak{K}_{g_1,0}$, where

$$f_1 = f - \Delta\psi, \quad g_1 = g - \psi.$$

Moreover, if one knows $v \in W^{2,p}_{\mathrm{loc}}(D)$ for some $p > n$, then it is also possible to show that

$$\Delta v = f_1 \chi_{\{v>0\}} \quad \text{in } D,$$

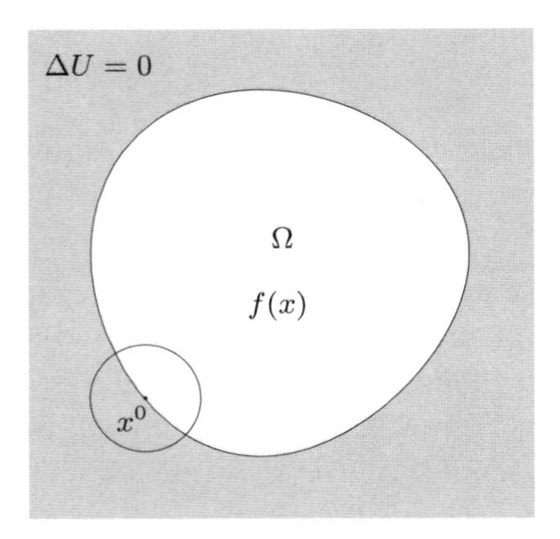

Figure 1.2. Harmonic continuation of Newtonian potentials

in the sense of distributions. This constitutes Exercise 1.2.

1.1.2. Problem from potential theory. Let Ω be a bounded open set in $\mathbb{R}^n$ and f a certain bounded measurable function on Ω. Consider then the Newtonian potential of the distribution of the mass $f\chi_\Omega$, i.e.

$$U(x) = \Phi_n * (f\chi_\Omega)(x) = \int_\Omega \Phi_n(x-y)f(y)dy,$$

where Φ_n is the fundamental solution of the Laplacian in $\mathbb{R}^n$, i.e. $\Delta\Phi_n = \delta$ in the sense of distributions. It can be shown (see e.g. Gilbarg-Trudinger [**GT01**, Theorem 9.9]) that the potential U is in $W^{2,p}_{\mathrm{loc}}(\mathbb{R}^n)$ for any $1 < p < \infty$ and satisfies

$$\Delta U = f\chi_\Omega \quad \text{in } \mathbb{R}^n,$$

in the sense of distributions (or a.e., which amounts to the same in this case). In particular, U is harmonic in $\mathbb{R}^n \setminus \overline{\Omega}$ (see Fig. 1.2).

Let $x^0 \in \partial\Omega$ and suppose for some small $r > 0$ there is a harmonic function h in the ball $B_r(x^0)$ such that $h = U$ on $B_r \setminus \Omega$. We say in this case that h is a harmonic continuation of U into Ω at x^0. If such a continuation exists, the difference $u = U - h$ satisfies

$$\begin{aligned} \Delta u &= f\chi_\Omega \quad \text{in } B_r(x^0), \\ u = |\nabla u| &= 0 \quad \text{on } B_r(x^0) \setminus \Omega. \end{aligned} \tag{1.4}$$

Using the Cauchy-Kovalevskaya theorem, it is straightforward to show that the harmonic continuation exists if $\partial\Omega$ and f are real-analytic in a neighborhood of x^0. One may ask the converse to this question: if U admits a harmonic continuation in a neighborhood of x^0, then what can be said about the regularity of $\partial\Omega$?

Note that when $U \geq h$, or equivalently $u \geq 0$, u solves the obstacle problem in $B_r(x^0)$ with zero obstacle. However, the sign condition on u is rather unnatural in this setting, which leads to many complications as compared to the obstacle problem. The problem (1.4) has also been called the *no-sign* obstacle problem in the literature.

We also note that the study of quadrature domains described in the Introduction (see p. 2) leads to the same problem (1.4) with $f \equiv 1$ locally near points on $\partial\Omega$.

1.1.3. Pompeiu problem. A nonempty bounded open set $\Omega \subset \mathbb{R}^n$ is said to have the Pompeiu property if the only continuous function such that

$$\int_{\sigma(\Omega)} f(x)dx = 0$$

for all rigid motions $\sigma : \mathbb{R}^n \to \mathbb{R}^n$ is the identically zero function. A ball of any radius R fails this property: take $f(x) = \sin(ax_1)$ for $a > 0$ satisfying $J_{n/2}(aR) = 0$, where J_ν is the Bessel function of order ν. Furthermore, any finite disjoint union of balls of the same radius again fail the Pompeiu property, with the same function f.

A long standing conjecture in integral geometry says that if Ω fails the Pompeiu property and has a sufficiently regular (Lipschitz) boundary $\partial\Omega$ homeomorphic to the unit sphere, then Ω must be a ball. It is known (see Williams [**Wil76**]) that for such Ω there exists a solution to the problem

$$\Delta u + \lambda u = \chi_\Omega \quad \text{in } \mathbb{R}^n, \qquad u = |\nabla u| = 0 \quad \text{in } \mathbb{R}^n \setminus \Omega,$$

for some $\lambda > 0$. This is a special case of an open conjecture of Schiffer which says that any Ω admitting solutions of the overdetermined problem above must be a ball.

1.1.4. A problem from superconductivity. In analyzing the evolution of vortices arising in the mean-field model of penetration of the magnetic field into superconducting bodies, one ends up with a degenerate parabolic-elliptic system. A simplified stationary model of this problem (in a local setting), where the scalar stream function admits a functional dependence on the scalar magnetic potential, reduces to finding u such that

$$\Delta u = f(x, u)\chi_{\{|\nabla u|>0\}} \quad \text{in } B_r(x^0),$$

with appropriate boundary conditions, where $f > 0$ and $f \in C^{0,1}(\mathbb{R}^n \times \mathbb{R})$.

This problem is more general than the problem in potential theory described in §1.1.2. For instance, the set $\Lambda = \{|\nabla u| = 0\}$ may consist of different components ("patches") Λ_j with u taking different constant values c_j on Λ_j (see Fig. 1.3). Exercise 1.4 at the end of this chapter gives an example of a solution with two different patches.

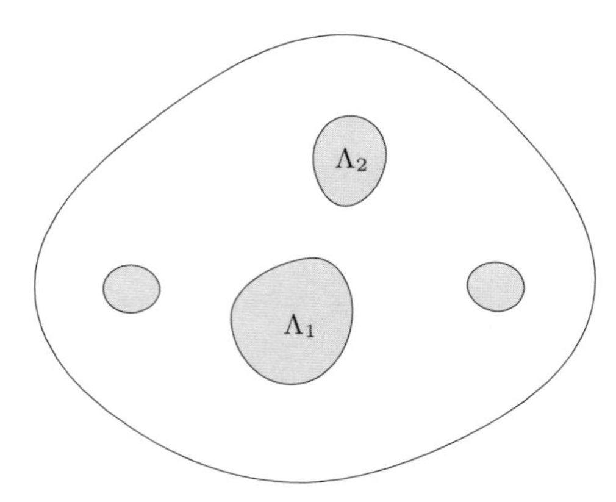

Figure 1.3. A problem from superconductivity: a solution with several "patches" $u \equiv c_j$ on Λ_j

1.1.5. Two-phase membrane problem. Given a bounded open set D in $\mathbb{R}^n$, $g \in W^{1,2}(D)$ and bounded measurable functions f_+ and f_- in D consider the problem of minimization of the functional

$$J_{f_+,f_-}(u) = \int_D (|\nabla u|^2 + 2f_+(x)u^+ + 2f_-(x)u^-)dx \tag{1.5}$$

over the set

$$\mathfrak{K}_g = \{u \in W^{1,2}(D) : u - g \in W_0^{1,2}(D)\}.$$

Here $u^+ = \max\{u, 0\}$, $u^- = \max\{-u, 0\}$. When $f_- = 0$ and $g \geq 0$, the problem is equivalent to the obstacle problem with zero obstacle discussed at the end of §1.1.1.

To give a physical interpretation of this problem, consider a thin membrane (film), which is fixed on the boundary of a given domain, and some part of the boundary data of this film is below the surface of a thick liquid, heavier than the film itself (see Fig. 1.4). Now the weight of the film produces a force downwards (call it f_+) on the part of the film which is above the liquid surface. On the other hand, the part in the liquid is pushed upwards by a force f_-, since the liquid is heavier than the film. The equilibrium state of the film is given by a minimization of the above-mentioned functional.

One of the difficulties one confronts in this problem is that the interface $\{u = 0\}$ consists in general of two parts: one where the gradient of u is nonzero and one where the gradient of u vanishes. Close to points of the latter part we expect the gradient of u to have linear growth. However, because of the decomposition into two different types of growth, it is not possible to derive a growth estimate by classical techniques.

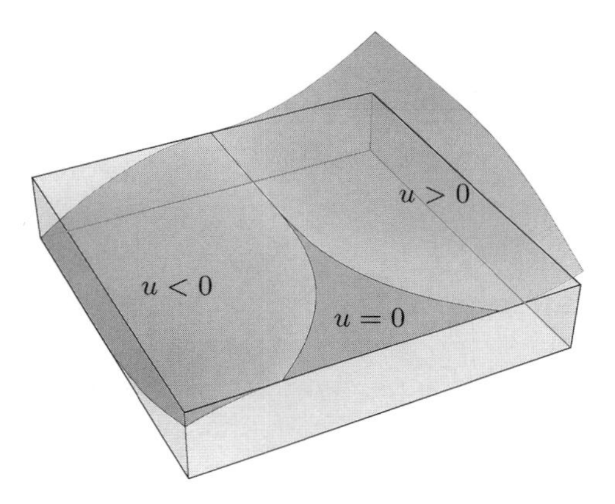

Figure 1.4. Two-phase membrane problem

Similarly to the classical obstacle problem, if one knows the $W^{2,p}_{\mathrm{loc}}$ regularity of the minimizer of J_{f_+,f_-} in D, then it will satisfy

$$\Delta u = f_+\chi_{\{u>0\}} - f_-\chi_{\{u<0\}} \quad \text{in } D.$$

This problem will be one of our model problems (Problem **B**) that we study in detail in this book, including the above-mentioned $W^{2,p}_{\mathrm{loc}}$ regularity, $1 < p < \infty$ (see §1.3.3).

1.1.6. Interior temperature control problem. Yet another application of the same equation as in the two-phase membrane problem arises in the temperature control through the interior, regulated by the temperature in the interior.

Suppose that we want to keep the temperature $u(x)$ in the domain D as close as possible to the range (θ_-, θ_+), where $\theta_-(x) \leq \theta_+(x)$ are two prescribed functions in D. We want to achieve this with the help of cooling/heating devices that are distributed evenly over the domain D. The devices are assumed to be of limited power, so the heat flux $-f$ generated by them will be in the range $[-f_-, f_+]$, $f_\pm \geq 0$. There will be no heat generated when $u(x) \in [\theta_-(x), \theta_+(x)]$; however, when u is not in that range, the corrective heat flux $-f$ will be injected according to the following rule:

$$-f = \Phi(u) = \begin{cases} \min\{k_+(u-\theta_+), f_+\}, & u > \theta_+, \\ \max\{k_-(u-\theta_-), -f_-\}, & u < \theta_-, \end{cases}$$

where $k_\pm \geq 0$. In the equilibrium state, the temperature distribution will satisfy

$$\Delta u = \Phi(u) \quad \text{in } D.$$

Assuming $\theta_- = \theta_+ = 0$ and $k_\pm = +\infty$, the equation becomes

$$\Delta u = f_+ \chi_{\{u>0\}} - f_- \chi_{\{u<0\}} \quad \text{in } D.$$

The model described above is from the book by Duvaut-Lions [**DL76**, Chapter I, §2.3.2].

1.1.7. Composite membrane. A free boundary problem of obstacle type appears also in the construction of composite membranes. One wants to build a body of prescribed shape consisting of given materials (of varying densities) in such a way that the body has a prescribed mass and that the basic frequency of the resulting membrane (with fixed boundary) is as small as possible. Let us consider a more general problem. Suppose we are given a domain $D \subset \mathbb{R}^n$ (bounded, connected, with Lipschitz boundary) and numbers $\alpha > 0$, $A \in [0, |D|]$. For any measurable subset $\Omega \subset D$ let $\lambda_D(\alpha, \Omega)$ denote the lowest eigenvalue λ of the problem

$$-\Delta v + \alpha \chi_\Omega v = \lambda v \quad \text{in } D, \qquad v = 0 \quad \text{on } \partial D, \tag{1.6}$$

and set

$$\Lambda_D(\alpha, A) := \inf_{\Omega \subset D,\ |\Omega| = A} \lambda_D(\alpha, \Omega).$$

Any minimizer Ω in the latter equation is called an optimal configuration for the data. If Ω is an optimal configuration and v satisfies (1.6), then (v, Ω) is called an optimal pair (or solution). It is known that $\Omega = \{v \le t\}$ for some t such that $A = |\{v \le t\}|$; see [**CGI⁺00**]. Now upon rewriting $u = v - t$ we can rephrase the above equation as

$$\Delta u = (\alpha \chi_{\{u \le 0\}} - \lambda)(u + t),$$

and with yet another rewriting we arrive at

$$\Delta u = \left((\alpha - \lambda)\chi_{\{u \le 0\}} - \lambda \chi_{\{u>0\}}\right)(u + t).$$

The particular case $\alpha < \lambda$ is of special interest, since the problem then does not fall into the category of obstacle-type problems, which we treat in detail in this book, but is rather akin to the *unstable obstacle problem*

$$\Delta u = -\chi_{\{u>0\}},$$

solutions of which fail in general to be $C^{1,1}$ (see the counterexample in §2.5).

1.1.8. Optimal stopping. In control theory one needs to find an optimal choice of strategy so that a cost/profit functional becomes minimal/maximal. Consider a domain $D \subset \mathbb{R}^n$, $x \in D$, and let $\mathbf{W}_x$ be a Brownian motion, with $\mathbf{W}_x(0) = x$. Further, let $\tau = \tau_x$ be a hitting time of ∂D and θ a stopping time (with respect to an underlying filtration for the Brownian motion).

For given functions f and g in $\overline{D}$, define an expected profit of stopping the Brownian motion at time $\theta \wedge \tau := \min(\theta, \tau)$ by

$$J_x(\theta) := \mathbb{E}\left[\int_0^{\theta\wedge\tau} \frac{1}{2} f(\mathbf{W}_x(s))ds + g(\mathbf{W}_x(\theta\wedge\tau))\right],$$

where $\mathbb{E}$ denotes the expected value. Now maximizing this over all possible stopping times and defining

$$u(x) := \sup_{\theta \text{ stopping time}} J_x(\theta),$$

it is relatively easy to show that u satisfies

$$u \geq g, \quad -\Delta u + f \geq 0, \quad (\Delta u - f)(u - g) = 0 \quad \text{in } D$$
$$u = g \quad \text{on } \partial D.$$

In other words, u solves the classical obstacle problem (1.2)–(1.3) with both the obstacle and boundary values given by g. For more explanation and details we refer to lecture notes of Evans [**Eva11**, pp. 103–107]. See also §9.1.2 for a version of this problem for jump processes.

1.1.9. Problems with lower-dimensional free boundaries. Over the past decade, there has been a renewed interest in problems that exhibit free boundaries of higher codimension, such as the thin obstacle problem (also known as the scalar Signorini problem), motivated by the new techniques as well as the increased range of applications.

On one hand, the methods for the study of such problems are quite similar to the ones for the problems described above, but on the other hand, there are substantial differences. For that reason, we decided to put the material related to the thin obstacle problem into a separate chapter, Chapter 9.

1.2. Model Problems A, B, C

Out of the examples described in the previous section, we have chosen three model problems (to be called Problems **A**, **B**, **C**) that we will study in detail throughout Chapters 2–8, as we believe they demonstrate the most typical techniques and difficulties associated with the obstacle-type problems. The forth model problem, Problem **S**, which is a version of the thin obstacle problem, will be treated in Chapter 9.

The model problems below have the following structure. We are given an open set D in $\mathbb{R}^n$ and a function $u \in L^\infty_{\mathrm{loc}}(D)$ that satisfies

$$\Delta u = f_u \quad \text{in} \quad D,$$

where f_u is a certain function in $L^\infty(D)$, having a jump discontinuity along a certain set $\Gamma(u)$, which is the *free boundary* in the problem. The last equation is understood in the sense of distributions, so that

$$\int_D u\Delta\eta\,dx = \int_D f_u\eta\,dx,$$

for all test functions $\eta \in C_0^\infty(D)$.

1.2.1. Problem A. *No-sign obstacle problem.* Our first model problem is the particular case of the problem from potential theory described in §1.1.2 with $f(x) \equiv 1$:

$$\textbf{(A)}\qquad \begin{aligned} \Delta u &= \chi_{\Omega(u)} \quad \text{in } D,\\ \Omega(u) &:= D \setminus \{u = |\nabla u| = 0\}. \end{aligned}$$

The free boundary in this case is $\Gamma(u) = \partial\Omega(u) \cap D$. As we noted earlier, if the solution u of this problem is nonnegative, then $\Omega = \{u > 0\}$ and u becomes a solution of the classical obstacle problem.

1.2.2. Problem B. *Superconductivity problem.* Our second model problem is the particular version of the problem from superconductivity in §1.1.4 with $f(x,u) \equiv 1$:

$$\textbf{(B)}\qquad \begin{aligned} \Delta u &= \chi_{\Omega(u)} \quad \text{in } D,\\ \Omega(u) &:= \{|\nabla u| > 0\}. \end{aligned}$$

The free boundary here is again $\Gamma(u) = \partial\Omega(u) \cap D$.

1.2.3. Problem C. *Two-phase membrane problem.* Our third model problem is the particular form of the problem in §1.1.5 with $f_\pm(x) \equiv \lambda_\pm$ positive constants:

$$\textbf{(C)}\qquad \begin{aligned} \Delta u &= \lambda_+\chi_{\Omega_+(u)} - \lambda_-\chi_{\Omega_-(u)} \quad \text{in } D,\\ \Omega_\pm(u) &:= \{\pm u > 0\}. \end{aligned}$$

The free boundary here is the union $\Gamma(u) = \Gamma_+(u) \cup \Gamma_-(u)$, where $\Gamma_\pm(u) = \partial\Omega_\pm(u) \cap D$. Since the behavior of u near the free boundary is going to be different depending on whether $|\nabla u|$ vanishes or not, we naturally subdivide Γ into the union of

$$\begin{aligned} \Gamma^0(u) &:= \Gamma(u) \cap \{|\nabla u| = 0\},\\ \Gamma^*(u) &:= \Gamma(u) \cap \{|\nabla u| \neq 0\}. \end{aligned}$$

Note that by the implicit function theorem Γ^* is locally a $C^{1,\alpha}$ graph (in fact, it can be shown to be real-analytic; see §4.5), so most of the difficulties lie in the study of the free boundary near the points in Γ^0.

1.2.4. Obstacle-type problems. Some of our results will be proved in a more general framework of what we will call *obstacle-type problems*, which is a slight abuse of the terminology, since there are still many interesting problems of obstacle type that do not fit into this framework.

We consider functions $u \in L^\infty_{\mathrm{loc}}(D)$ that satisfy

$$(\mathbf{OT}_1) \qquad \begin{aligned} \Delta u &= f(x,u)\chi_{G(u)} \quad \text{in } D,\\ |\nabla u| &= 0 \quad \text{on } D \setminus G(u), \end{aligned}$$

where $G(u)$ is an open subset of D and $f : D \times \mathbb{R} \to \mathbb{R}$ satisfies the following structural conditions: there exist $M_1, M_2 \geq 0$ such that

$$(\mathbf{OT}_2) \qquad \begin{aligned} |f(x,t) - f(y,t)| &\leq M_1|x-y|, \quad x, y \in D,\ t \in \mathbb{R},\\ f(x,s) - f(x,t) &\geq -M_2(s-t), \quad x \in D,\ s,t \in \mathbb{R},\ s \geq t. \end{aligned}$$

Locally, these conditions are equivalent to

$$|\nabla_x f(x,t)| \leq M_1, \quad \partial_t f(x,t) \geq -M_2$$

in the sense of distributions. The free boundary is going to be $\partial G(u) \cap D$ and/or the set of discontinuity of $f(x,u)$, depending on the problem.

Problems **A**, **B**, and **C** fit into the framework $\mathbf{OT}_1$–$\mathbf{OT}_2$ as follows:

- Problems **A**, **B**: $G = \Omega(u)$, $f(x,u) = 1$ in G. More generally, the problems $\Delta u = f(x)\chi_{\Omega(u)}$ with $\Omega(u)$ as in Problem **A** or **B** will fit in our framework if $f \in C^{0,1}(D)$.
- Problem **C**: $G = D$, $f(x,u) = \lambda_+\chi_{\Omega_+(u)} - \lambda_-\chi_{\Omega_-(u)}$. More generally, the equation $\Delta u = f_+(x)\chi_{\Omega_+(u)} - f_-(x)\chi_{\Omega_-(u)}$ fits in our framework if $f_\pm \in C^{0,1}(D)$.

1.3. $W^{2,p}$ regularity of solutions

1.3.1. Calderón-Zygmund estimates. In Chapter 2 we will see that the solutions u of obstacle-type problems $\mathbf{OT}_1$-$\mathbf{OT}_2$ enjoy the $C^{1,1}$ regularity, which is the optimal regularity of the solutions, since Δu is generally a discontinuous function. However, that will require the use of a rather powerful Alt-Caffarelli-Friedman monotonicity formula and its generalizations (see §2.2). Here, we remark that the almost optimal $W^{2,p}_{\mathrm{loc}} \cap C^{1,\alpha}_{\mathrm{loc}}$ regularity for any $1 < p < \infty$ and $0 < \alpha < 1$ comes "for free" from the standard L^p-theory of elliptic equations (see Gilbarg-Trudinger [**GT01**, Chapter 9]).

Theorem 1.1. *Let $u \in L^1(D)$, $f \in L^p(D)$, $1 < p < \infty$, be such that $\Delta u = f$ in D in the sense of distributions. Then $u \in W^{2,p}_{\mathrm{loc}}(D)$ and*

$$\|u\|_{W^{2,p}(K)} \leq C\left(\|u\|_{L^1(D)} + \|f\|_{L^p(D)}\right),$$

for any $K \Subset D$ with $C = C(p,n,K,D)$. □

Remark 1.2. It should be noted that in the typical versions of Theorem 1.1 in the literature one has $\|u\|_{L^p(D)}$ instead of $\|u\|_{L^1(D)}$ on the right-hand side and a priori assumes that $u \in W^{2,p}_{\mathrm{loc}}(D) \cap L^p(D)$; see e.g. Gilbarg-Trudinger [**GT01**, Theorem 9.11]. However, that is a superfluous assumption. The proof is outlined in Exercise 1.6, based on the $W^{2,p}$ estimates for the Newtonian potentials.

Thus, for solutions of $\Delta u = f$ in D with $u \in L^1(D)$ and $f \in L^\infty(D)$ we have

$$u \in W^{2,p}_{\mathrm{loc}}(D), \quad \text{for all} \quad 1 < p < \infty. \tag{1.7}$$

Consequently, we also have

$$u \in C^{1,\alpha}_{\mathrm{loc}}(D), \quad \text{for all} \quad 0 < \alpha < 1, \tag{1.8}$$

by the Sobolev embedding $W^{2,p} \hookrightarrow C^{1,\alpha}$ with $\alpha = 1 - n/p$ for $p > n$.

An easy counterexample (see Exercise 1.7) shows that in general we cannot have $p = \infty$ in (1.7) and $\alpha = 1$ in (1.8). Instead, we have the following estimate.

Theorem 1.3. *Let $u \in L^1(D)$, $f \in L^\infty(D)$ be such that $\Delta u = f$ in the sense of distributions. Then $u \in W^{2,p}_{\mathrm{loc}}(D) \cap C^{1,\alpha}_{\mathrm{loc}}(D)$ for all $1 < p < \infty$, $0 < \alpha < 1$ and*

$$|\nabla u(x) - \nabla u(y)| \leq C(\|u\|_{L^1(D)} + \|f\|_{L^\infty(D)})|x - y| \log \frac{1}{|x - y|},$$

for any $x, y \in K \Subset D$ with $|x - y| \leq 1/e$ and $C = C(n, K, D)$. □

We refer again to Exercise 1.6 for the outline of the proof.

As we will see later, the logarithmic term in this theorem can be dropped if one assumes the additional structure on the right-hand side as in $\mathbf{OT}_1$–$\mathbf{OT}_2$. That would give us a starting point for the analysis of the free boundary in those problems.

1.3.2. Classical obstacle problem.

The $W^{2,p}$ estimates come "for free" in Problems **A**, **B**, and **C**, since we state them in the form of equations with bounded right-hand sides. However, recall that to obtain such an equation for the solutions of the classical obstacle problem (§1.1.1) or the two-phase membrane problem (§1.1.5), we needed to have the $W^{2,p}$ regularity in the first place. So to avoid this circular reasoning, we establish the $W^{2,p}$ regularity here.

We start with the classical obstacle problem. Recall that by simply subtracting the obstacle (see §1.1.1.3) we reduce it to the case of the obstacle problem with zero obstacle, i.e. the problem of minimizing

$$J(u) = \int_D (|\nabla u|^2 + 2fu)dx$$

over the convex closed subset

$$\mathfrak{K}_{g,0} = \{u \in W^{1,2}(D) : u - g \in W_0^{1,2}(D), u \geq 0 \text{ a.e. in } D\}.$$

Here we assume $g \in W^{1,2}(D)$, $g \geq 0$ on ∂D in the sense that $g^- \in W_0^{1,2}(D)$ and $f \in L^\infty(D)$.

It is easy to see that the functional J is strictly convex and bounded below on $\mathfrak{K}_{g,0}$. Hence, J has a unique minimizer on $\mathfrak{K}_{g,0}$. We show next that the minimizer is in $W^{2,p}_{\mathrm{loc}}(D)$ for any $1 < p < \infty$ and will consequently solve

$$\Delta u = f\chi_{\{u>0\}} \quad \text{in } D.$$

The first step is getting rid of the obstacle at the expense of losing the regularity of the functional J.

Lemma 1.4. *A function $u \in W^{1,2}(D)$ is a minimizer of J over $\mathfrak{K}_{g,0}$ iff u is a minimizer of the functional*

$$\tilde{J}(u) = \int_D (|\nabla u|^2 + 2fu^+)dx$$

over $\mathfrak{K}_g = \{u \in W^{1,2}(D) : u - g \in W_0^{1,2}(D)\}$.

Proof. For any $u \in \mathfrak{K}_g$ we have that $u^+ \in \mathfrak{K}_{g,0}$ and

$$\nabla(u^+) = (\nabla u)\chi_{\{u>0\}}.$$

We then claim that

$$\tilde{J}(u^+) \leq \tilde{J}(u) \quad \text{for any } u \in \mathfrak{K}_g,$$

with equality iff $u = u^+$. Indeed,

$$\tilde{J}(u) = \int_D (|\nabla u|^2 + 2fu^+)dx \geq \int_D (|\nabla u|^2\chi_{\{u>0\}} + 2fu^+)dx = \tilde{J}(u^+).$$

The equality $\tilde{J}(u^+) = \tilde{J}(u)$ holds iff $\nabla u = 0$ a.e. on $\{u \leq 0\}$, which is equivalent to having $\nabla u^- = 0$ a.e. in D. The latter means that u^- is locally constant, and since $u^- \in W_0^{1,2}(D)$, it follows that $u^- = 0$. Hence, $\tilde{J}(u^+) = \tilde{J}(u)$ iff $u \geq 0$ a.e. in D, or equivalently $u \in \mathfrak{K}_{g,0}$. Thus, $\tilde{J}$ attains its minimum on $\mathfrak{K}_{g,0}$.

On the other hand

$$\tilde{J} = J \quad \text{on } \mathfrak{K}_{g,0}.$$

Hence these two functionals have the same set of minimizers. □

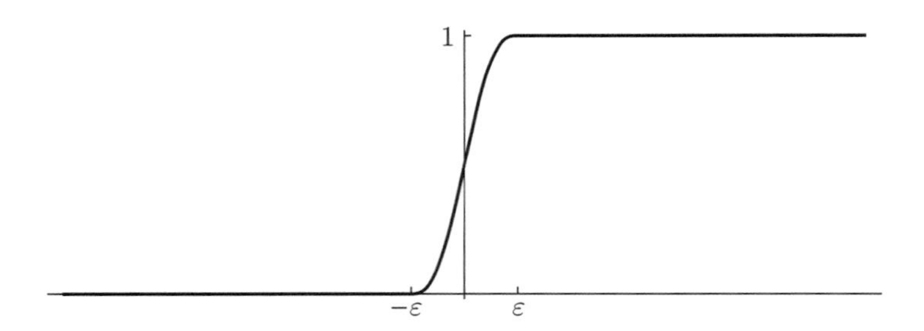

Figure 1.5. The function χ_ε

Thus, we reduced the problem to studying the minimizers of $\tilde{J}$ with given boundary values g on ∂D. In the next step, consider a family of regularized problems

$$\begin{aligned} \Delta u_\varepsilon &= f\chi_\varepsilon(u_\varepsilon) && \text{in } D,\\ u_\varepsilon &= g && \text{on } \partial D, \end{aligned}$$

for $0 < \varepsilon < 1$, where $\chi_\varepsilon(s)$ is a smooth approximation of the Heaviside function $\chi_{\{s>0\}}$ such that

$$\chi_\varepsilon' \geq 0, \quad \chi_\varepsilon(s) = 0 \quad \text{for } s \leq -\varepsilon, \quad \chi_\varepsilon(s) = 1 \quad \text{for } s \geq \varepsilon.$$

A solution u_ε to this problem can be obtained by minimizing the functional

$$J_\varepsilon(u) = \int_D (|\nabla u|^2 + 2f(x)\Phi_\varepsilon(u))dx$$

over $\mathfrak{K}_g$, where

$$\Phi_\varepsilon(s) = \int_{-\infty}^s \chi_\varepsilon(t)dt.$$

To be more precise, the minimizer u_ε will satisfy $\Delta u_\varepsilon = f(x)\chi_\varepsilon(u_\varepsilon)$ in D in the sense that

$$\int_D (\nabla u_\varepsilon \nabla \eta + f(x)\chi_\varepsilon(u_\varepsilon)\eta) = 0$$

for any $\eta \in W_0^{1,2}(D)$. We next want to show that the family $\{u_\varepsilon\}$ is uniformly bounded in $W^{1,2}(D)$ as well as in $W^{2,p}(K)$ for any $K \Subset D$.

To show the uniform $W^{1,2}(D)$ estimate, we take $\eta = u_\varepsilon - g$ in the above equation. Then an application of the Poincaré inequality gives

$$\int_D |\nabla(u_\varepsilon - g)|^2 dx \leq C(f, g),$$

uniformly for $0 < \varepsilon < 1$. This implies the claimed uniform boundedness of $\{u_\varepsilon\}$ in $W^{1,2}(D)$.

As a consequence, we obtain that there exists $u \in W^{1,2}(D)$ such that over a subsequence $\varepsilon = \varepsilon_k \to 0$,

$$\begin{aligned} u_\varepsilon &\to u \quad \text{weakly in } W^{1,2}(D), \\ u_\varepsilon &\to u \quad \text{strongly in } L^2(D). \end{aligned}$$

Moreover, since $u_\varepsilon - g \in W^{1,2}_0(D)$ and $W^{1,2}_0(D)$ is a closed subspace (thus also closed with respect to weak convergence) of the Hilbert space $W^{1,2}(D)$, we conclude that $u \in \mathfrak{K}_g$.

Next, recalling that u_ε are weak solutions of the regularized problem $\Delta u_\varepsilon = f(x)\chi_\varepsilon(u_\varepsilon)$ and applying the Calderón-Zygmund estimate (Theorem 1.1), we will obtain that

$$\begin{aligned} \|u_\varepsilon\|_{W^{2,p}(K)} &\leq C(K,D)(\|u_\varepsilon\|_{L^2(D)} + \|f\chi_\varepsilon(u_\varepsilon)\|_{L^\infty(D)}) \\ &\leq C(K,D,f,g), \end{aligned}$$

for any $K \Subset D$ and $1 < p < \infty$. Thus, we may further assume that over the same sequence $\varepsilon = \varepsilon_k \to 0$ as above,

$$u_\varepsilon \to u \quad \text{weakly in } W^{2,p}_{\text{loc}}(D),$$

for any $1 < p < \infty$. Clearly, $u \in W^{2,p}_{\text{loc}}(D)$ for any $1 < p < \infty$.

Now, we are ready to pass to the limit as $\varepsilon = \varepsilon_k \to 0$. Since

$$\begin{aligned} \int_D |\nabla u|^2 dx &\leq \liminf_{\varepsilon \to 0} \int_D |\nabla u_\varepsilon|^2 dx, \\ \int_D f u^+ dx &= \lim_{\varepsilon \to 0} \int_D f \Phi_\varepsilon(u_\varepsilon) dx, \end{aligned}$$

we see that

$$\tilde{J}(u) \leq \liminf_{\varepsilon = \varepsilon_k \to 0} J_\varepsilon(u_\varepsilon) \leq \liminf_{\varepsilon = \varepsilon_k \to 0} J_\varepsilon(v) = \tilde{J}(v),$$

for any $v \in \mathfrak{K}_g$. Thus, by Lemma 1.4, u is the solution of the obstacle problem. Finally, we verify that u satisfies

$$\Delta u = f\chi_{\{u>0\}} \quad \text{in } D,$$

in the sense of distributions. Since $u \in W^{2,p}_{\text{loc}}(D)$, we readily have that Δu is an L^p function locally in D, and thus we have to verify that

$$\Delta u = f\chi_{\{u>0\}} \quad \text{a.e. in } D.$$

To this end, we remark that since $u_\varepsilon \in W^{2,p}_{\text{loc}}(D)$, the equation $\Delta u_\varepsilon = f(x)\chi_\varepsilon(u_\varepsilon)$ is now satisfied in the strong sense, i.e., for a.e. $x \in D$. Moreover, taking p large, by the Sobolev embedding theorem we may assume that over $\varepsilon = \varepsilon_k \to 0$ we have

$$u_\varepsilon \to u \quad \text{in } C^{1,\alpha}_{\text{loc}}(D).$$

Then the locally uniform convergence implies that $\Delta u = f$ a.e. in the open set $\{u > 0\}$. Besides, using the fact that $u \in W^{2,p}_{\mathrm{loc}}(D)$ one more time, we get $\Delta u = 0$ a.e. on $\{u = 0\}$. Combining these two facts, we obtain that $\Delta u = f(x)\chi_{\{u>0\}}$ a.e. in D.

1.3.3. Two-phase membrane problem. Given any two nonnegative bounded functions f_+ and f_- we want to show here that the minimizer of the energy functional

$$J(v) = \int_D (|\nabla v|^2 + 2f_+u^+ + 2f_-u^-)dx$$

on the set $\mathfrak{K}_g$, $g \in W^{1,2}(D)$, is in $W^{2,p}_{\mathrm{loc}}(D)$ for any $1 < p < \infty$ and solves

$$\Delta u = f_+(x)\chi_{\{u>0\}} - f_-(x)\chi_{\{u<0\}} \quad \text{a.e. in } D.$$

To this end consider the approximating problems

$$\begin{aligned} \Delta u &= f_+\chi_\varepsilon(u) - f_-\chi_\varepsilon(-u) && \text{in } D,\\ u &= g && \text{on } \partial D, \end{aligned}$$

and the solutions u_ε obtained by minimizing the functional

$$J_\varepsilon(u) = \int_D (|\nabla u|^2 + 2f_+\Phi_\varepsilon(u) + 2f_-\Phi_\varepsilon(-u))dx,$$

where the approximations χ_ε and Φ_ε are as in the previous subsection. Then, following the arguments as before one can establish that for a subsequence $\varepsilon = \varepsilon_k \to 0$ the minimizers u_ε converge weakly in $W^{2,p}_{\mathrm{loc}}(D)$ for any $1 < p < \infty$ to a solution of the desired problem. We leave it to the reader to fill in the details (see Exercise 1.9).

Notes

The obstacle problem originated in the work of Stampacchia [**Sta64**], with the obstacle being a characteristic function of a set (in relation to the capacity of that set). The Signorini problem (akin to the thin obstacle problem; see Chapter 9) have appeared even earlier in Signorini [**Sig59**] and Fichera [**Fic64**]. Lions-Stampacchia [**LS67**] gave the first systematic treatment of variational inequalities (of which the obstacle and Signorini problems are particular examples).

Problems from potential theory similar to the one in §1.1.2 and Problem **A** can be found in Sakai [**Sak82, Sak91, Sak93**]. We also refer to the lectures of Gustafsson [**Gus04**] for related problems. This kind of problems are also of interest in inverse problems and geophysics; see e.g. Strakhov [**Str74**], Margulis [**Mar82**], and the book by Isakov [**Isa90**]. Our main reference for Problem **A** is Caffarelli-Karp-Shahgholian [**CKS00**].

The Pompeiu problem in §1.1.3 originated in the work of Pompeiu [**Pom29**]. The problem was reduced by Williams [**Wil76**] to a special case in Schiffer's conjecture (see e.g. Yau [**Yau82**, p. 688, Problem 80]). Analyticity of the boundary of the domains failing the Pompeiu property was proved by Williams [**Wil81**] in case the boundary is assumed to be Lipschitz. This was extended to domains satisfying a thickness condition by Caffarelli-Karp-Shahgholian [**CKS00**].

The problem from superconductivity in §1.1.4 (and also Problem **B**) is a simplified time-independent version of the parabolic-elliptic mean-field model of Chapman [**Cha95**]. There are related models by Berestycki-Bonnet-Chapman [**BBC94**] and Chapman-Rubinstein-Schatzman [**CRS96**], with a rigorous derivation from the Ginzburg-Landau model by Sandier-Serfaty [**SS00**]. Viscosity solutions were studied by Elliott-Schätzle-Stoth [**ESS98**] and Caffarelli-Salazar [**CS02**]. For particular configurations (with single patches) the free boundary was studied by Bonnet-Monneau [**BM00**] and Monneau [**Mon04**]. The free boundary in the general case was first studied by Caffarelli-Salazar [**CS02**], followed by Caffarelli-Salazar-Shahgholian [**CSS04**]. The latter paper is our main source for the results on Problem **B**.

The two-phase membrane problem in §1.1.5 (and Problem **C**) was known in the literature at least since the mid 1970s as a limiting case in the interior temperature control problem, as described in §1.1.6 (see also Duvaut-Lions [**DL76**, Chapter I, §2.3.2]). From the free boundary point of view it was first studied by Weiss [**Wei01**]. There is now more or less complete understanding of the problem, thanks to the works by Uraltseva [**Ura01**], Shahgholian-Weiss [**SW06**], and Shahgholian-Uraltseva-Weiss [**SUW07**].

The composite membrane problem in §1.1.7 can be found in Chanillo-Grieser-Imai-Kurata-Ohnishi [**CGI^{+}00**]. Partial regularity of the free boundary was proved in dimension two by Shahgholian [**Sha07**] (see also Blank [**Bla04**]) and Chanillo-Kenig [**CK08**] in higher dimensions. The full regularity (analyticity) in dimension two was proved by Chanillo-Kenig-To [**CKT08**]. The composite membrane problem is related to the unstable obstacle problem, first studied by Monneau-Weiss [**MW07**] in connection with a model in solid combustion. For some recent results on this interesting problem, we refer to Andersson-Shahgholian-Weiss [**ASW10**].

The optimal stopping problems in §1.1.8 can be viewed as a simplified version of the optimal pricing problem in mathematical finance for the so-called American (call/put) options, see e.g. Wilmott-Howison-Dewynne [**WHD95**, Chap. 7] and Evans [**Eva98**, pp. 107–111]. Optimal stopping problems for jump processes (stable Lévy processes) were also studied in the literature. They lead to obstacle-type problems for nonlocal integro-differential operators; see e.g. the thesis of Silvestre [**Sil07**]. In a particular case of the half-Laplacian, this has a direct relation to the thin obstacle problem; see §9.1.2 for more details.

The $W^{2,p}$ regularity for solutions of the obstacle problem is due to Lewy-Stampacchia [**LS69, LS70, LS71**]. The method of penalization (as in Exercise 1.10) was used in Brezis-Stampacchia [**BS68**], Lions [**Lio69**], Brezis [**Bre72**], and Brezis-Kinderlehrer [**BK74**].

In the recent literature, the obstacle problem was generalized to governing operators of various types. Below, we mention just a few of the available papers. Obstacle-type problems for p-Laplacian were studied by Choe-Lewis [**CL91**], Karp-Kilpeläinen-Petrosyan-Shahgholian [**KKPS00**], and Lee-Shahgholian [**LS03**]; see also the references therein. Obstacle-type problems for uniformly elliptic fully nonlinear equations were studied by Lee in [**Lee98**] and for Monge-Ampére equation in [**Lee01**]. A subelliptic obstacle problem was studied by Danielli-Garofalo-Salsa [**DGS03**] and Danielli-Garofalo-Petrosyan [**DGP07**].

Exercises

1.1. Let u be a solution of the obstacle problem as in §1.1.1.2. Under the additional assumption that $u \in W^{2,p}_{\mathrm{loc}}(D)$ for some $p > n$ prove that

$$\begin{aligned} \Delta u &= f && \text{in } \Omega = \{u > \psi\}, \\ \Delta u &= \Delta\psi && \text{a.e. on } \Lambda = \{u = \psi\}, \end{aligned}$$

and that

$$\Delta u \leq f \qquad \text{in } D,$$

in the sense of distributions, or equivalently,

$$\int_D (\nabla u \nabla \eta + f\eta)dx \geq 0,$$

for any nonnegative $\eta \in W^{1,2}_0(D)$. Verify that this implies the complementarity condition (1.2).

Hint: To show that $\Delta u \leq f$ let $\varepsilon \to 0+$ in

$$\frac{J(u+\varepsilon\eta) - J(u)}{\varepsilon} \geq 0.$$

1.2. Prove the assertions in §1.1.1.3, i.e. the reduction of the classical obstacle problem to the case with zero obstacle.

1.3. Show that in §1.1.2, if $\partial\Omega \cap B_r(x^0)$ is a real-analytic surface for some $r > 0$, then the Newtonian potential U admits a harmonic continuation into Ω at x^0. Use the Cauchy-Kovalevskaya theorem.

1.4. The following example constructs a solution to the problem in §1.1.4 with two different "patches" of the set $\{|\nabla u| = 0\}$.

Consider a dumbbell-shaped region $D \subset \mathbb{R}^2$,

$$D := B_1(x^1) \cup B_1(x^2) \cup \{x : |x_2| < \varepsilon,\ |x_1| < 2\}$$

with $x^1 = (2,0)$, $x^2 = (-2,0)$, and ε very small positive number. The solution to

$$\Delta v = f \quad \text{in } D, \qquad v = 0 \quad \text{on } \partial D,$$

forms a shape of hanging graphs over D, symmetric with respect to the x_1-axis.

Then solve the obstacle problem $\Delta u = f\chi_{\{u>\psi\}}$ in D with zero boundary values and the obstacle ψ which is smooth and equal to $\min v + \delta_i$ on each ball $B_{1/2}(x^i)$. Here $\delta_1 > \delta_2 > 0$ are small constants.

1.5. Let D be an open set with a Lipschitz boundary in $\mathbb{R}^n$. For a given function $h \in L^\infty(\partial D)$ and a control function $f \in L^\infty(D)$ from the class

$$U_{\text{ad}} = \left\{ \sup_D |f| \le 1, \int_D f = \int_{\partial D} h \right\},$$

consider the following problem:

$$\Delta u = f \quad \text{in } D, \quad \partial_\nu u = h \quad \text{on } \partial D.$$

Here ν is an outward normal to ∂D. Then minimize the functional

$$I(u) = \int_D |\nabla u|^2 + |u| - \int_{\partial D} hu$$

among all solutions with $f \in U_{\text{ad}}$. Show that the minimizer satisfies

$$\Delta u = \operatorname{sgn} u \quad \text{in } D,$$

in the weak sense. Note that this is a particular case of the two-membrane problem in §1.1.5 with $f_\pm \equiv 1$.

Hint: Show that for any $f \in U_{\text{ad}}$,

$$I(u) = \int_D |u|(1 - f \operatorname{sgn} u).$$

Consequently, $I(u) \ge 0$ for solutions u for any $f \in U_{\text{ad}}$, and the minimum $I(u) = 0$ is attained if $f = \operatorname{sgn} u$.

1.6. (i) Let D be a bounded open set in $\mathbb{R}^n$, $n \ge 3$, $f \in L^p(D)$, $1 < p < \infty$. Consider the Newtonian potential

$$w(x) = c_n \int_D f(y)|x - y|^{2-n} dy,$$

with the appropriately chosen c_n. Use the $W^{2,p}$ estimates for w (see [**GT01**, Lemma 7.12, Theorem 9.9]):

$$\|w\|_{L^p(D)} + \|D^2 w\|_{L^p(D)} \le C(n, D)\|f\|_{L^p(D)}$$

to deduce Theorem 1.1.

Hint: Represent $u = v + w$, where w is the Newtonian potential as before and v a harmonic function in D. Then use the estimates above combined with the interior derivative estimates for harmonic functions (see [**Eva98**, §2.2, Theorem 7]),

$$\|D^k v\|_{L^\infty(K)} \le C\|v\|_{L^1(D)},$$

to deduce Theorem 1.1. In $\mathbb{R}^2$, add a dummy variable to extend the functions to $\mathbb{R}^3$.

(ii) Assume now that $f \in L^\infty(D)$ and w is the Newtonian potential of f as above. Then use the estimate (see [**Mor08**, Theorem 2.5.1])

$$|\nabla w(x) - \nabla w(y)| \le C(n, K, D)\|f\|_{L^\infty(D)}|x - y| \log \frac{1}{|x - y|}$$

for $x, y, \in K \Subset D$, $|x - y| < 1/e$, to deduce Theorem 1.3.

1.7. Show that the function $u(x_1, x_2) = (x_1^2 - x_2^2)\log(x_1^2 + x_2^2)$ defined in $\mathbb{R}^2$ is locally bounded and satisfies $\Delta u = f$ in the sense of distributions for a certain $f \in L^\infty(\mathbb{R}^2)$ but is not in $W^{2,\infty}_{\mathrm{loc}}(\mathbb{R}^2) = C^{1,1}_{\mathrm{loc}}(\mathbb{R}^2)$.

1.8. Check that to prove the $W^{2,p}_{\mathrm{loc}}$ estimates for the minimizers of the functional J in §1.3.2 for a fixed $1 < p < \infty$, it is enough to assume that $f \in L^p(D)$. [Showing that the minimizer u solves $\Delta u = f\chi_{\{u>0\}}$ may not be easy if $p < n/2$, since we do not even know if the set $\{u > 0\}$ is open in that case.]

1.9. Complete the proof of the $W^{2,p}_{\mathrm{loc}}$ regularity of the solutions of the two-phase membrane problem in §1.3.3.

1.10. The purpose of this exercise is to give an alternative proof of the $W^{2,p}_{\mathrm{loc}}$ regularity of solutions of the obstacle problem. This is known as the *method of penalization.* We will assume that ∂D is sufficiently smooth and that the boundary data $g \in C^{1,\alpha}(\partial D)$ for some $\alpha > 0$.

For $\varepsilon > 0$, let $\beta_\varepsilon \in C^\infty(\mathbb{R})$ be such that

$$\beta_\varepsilon' \ge 0, \quad \beta_\varepsilon \le 0, \quad \beta_\varepsilon(s) = 0 \text{ for } s \ge 0, \quad \beta_\varepsilon(s) = \frac{s}{\varepsilon} \text{ for } s < -\varepsilon.$$

Also let $\beta_{\varepsilon,N} = \max\{\beta_\varepsilon, -N\}$ for any $N > 0$. Now let $u^{\varepsilon,N}$ be a solution of the *penalized problem*

$$\begin{aligned} \Delta u &= \beta_{\varepsilon,N}(u) + f(x) \quad \text{in } D,\\ u &= g \quad \text{on } \partial D. \end{aligned}$$

(a) Prove that the function $\zeta^{\varepsilon,N}(x) = \beta_{\varepsilon,N}(u^{\varepsilon,N}(x))$ is bounded in D, uniformly in ε and N. In fact,

$$-\|f\|_{L^\infty(D)} \le \zeta^{\varepsilon,N} \le 0 \quad \text{in } D.$$

Hint: Let x^0 be a point where $\zeta^{\varepsilon,N}$ achieves its negative minimum. Then $x^0 \in D$ and $u^{\varepsilon,N}$ has a local minimum at that point, which implies that $\Delta u^{\varepsilon,N}(x^0) \le 0$. Use this fact to find a bound on $\zeta^{\varepsilon,N}(x^0)$ from below.

(b) Conclude from (a) that $u^\varepsilon = u^{\varepsilon,N}$ for large enough N solves

$$\begin{aligned} \Delta u = \beta_\varepsilon(u) + f(x) &\quad \text{in } D, \\ u = g &\quad \text{on } \partial D, \end{aligned}$$

and that the u^ε are uniformly bounded in $W^{2,p}(K)$ for $K \Subset D$, and in $C^{1,\alpha}(\overline{D})$ for some $\alpha > 0$.

(c) Rigorously show that the limit u of any subsequence u^{ε_k} with $\varepsilon_k \to 0$ is a solution of the obstacle problem with zero obstacle and boundary data g.
Hint: Use (a) to deduce that $u \ge 0$ in D and show that the complementarity condition (1.2) holds with $\psi = 0$.

Chapter 2

Optimal regularity of solutions

The purpose of this chapter is to establish the $C^{1,1}$ regularity of solutions of Problems **A**, **B**, and **C**. In fact, we will do that for a more general class of functions solving $\mathbf{OT}_1$–$\mathbf{OT}_2$ (§2.3), with the help of Alt-Caffarelli-Friedman monotonicity formula and its generalizations (§2.2). We then show that the estimates can be extended up to flat boundaries with zero Dirichlet data (§2.4). We also show the importance of the structural conditions $\mathbf{OT}_1$–$\mathbf{OT}_2$ by constructing a non-$C^{1,1}$ solution of the so-called unstable obstacle problem (§2.5).

We start, however, with the proof of $C^{1,1}$ regularity in the classical obstacle problem, which is much simpler and does not require monotonicity formulas, but is rather based on the maximum principle and Harnack inequality (§2.1).

2.1. Optimal regularity in the classical obstacle problem

Let $u \in L^\infty(D)$ be a nonnegative solution of

$$\Delta u = f(x)\chi_{\{u>0\}} \quad \text{in } D, \tag{2.1}$$

for $f \in L^\infty(D)$. While we already know that $u \in W^{2,p}_{\mathrm{loc}}(D)$ for any $1 < p < \infty$ (see §1.3.2), we will show that in fact $u \in C^{1,1}_{\mathrm{loc}}(D)$ (under additional natural conditions on f).

The first step is the following estimate on the growth of u away from the free boundary $\Gamma(u) = \partial\Omega(u) \cap D$, $\Omega(u) = \{u > 0\}$.

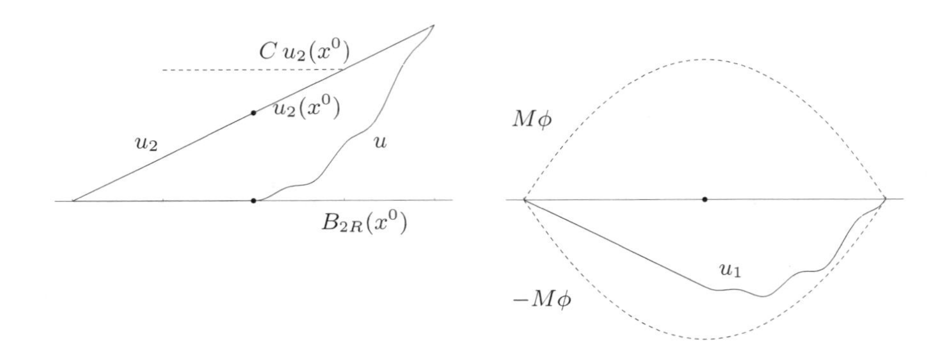

Figure 2.1. Splitting u into the sum $u_1 + u_2$

Theorem 2.1 (Quadratic growth). *Let $u \in L^\infty(D)$, $u \geq 0$, satisfy (2.1), $x^0 \in \Gamma(u)$, and $B_{2R}(x^0) \subset D$. Then*

$$\sup_{B_R(x^0)} u \leq C_n \|f\|_{L^\infty(D)} R^2.$$

Proof. Split u into the sum $u_1 + u_2$ in $B_{2R}(x^0)$, where

$$\begin{aligned} \Delta u_1 &= \Delta u, \quad \Delta u_2 = 0 \quad \text{in } B_{2R}(x^0),\\ u_1 &= 0, \quad u_2 = u \quad \text{on } \partial B_{2R}(x^0). \end{aligned}$$

We then estimate the functions u_1 and u_2 separately (see Fig. 2.1).

1) To estimate u_1, we consider the auxiliary function

$$\phi(x) = \frac{1}{2n}(4R^2 - |x - x^0|^2),$$

which is the solution of

$$\Delta \phi = -1 \quad \text{in } B_{2R}(x^0), \qquad \phi = 0 \quad \text{on } \partial B_{2R}(x^0).$$

Then we have

$$-M\phi(x) \leq u_1(x) \leq M\phi(x), \quad x \in B_{2R}(x^0),$$

where $M = \|f\|_{L^\infty(D)}$. This follows from the comparison principle, since

$$-M \leq \Delta u_1 \leq M \quad \text{in} \quad B_{2R}(x^0)$$

and both u_1 and ϕ vanish on $\partial B_{2R}(x^0)$. In particular, this implies that

$$|u_1(x)| \leq \frac{2}{n} M R^2, \quad x \in B_{2R}(x^0).$$

2) To estimate u_2, we observe that $u_2 \geq 0$ in $B_{2R}(x^0)$, since u_2 is harmonic in $B_{2R}(x^0)$ and $u_2 = u \geq 0$ on $\partial B_{2R}(x^0)$. Also note that $u_1(x^0) + u_2(x^0) = u(x^0) = 0$ and the estimate of u_1 gives

$$u_2(x^0) = -u_1(x^0) \leq \frac{2}{n} MR^2.$$

Applying now the Harnack inequality, we obtain

$$u_2(x) \leq C_n u_2(x^0) \leq C_n MR^2, \quad x \in B_R(x^0).$$

Finally, combining the estimates for u_1 and u_2, we obtain the desired estimate for u. □

Corollary 2.2. *Let u be as in Theorem 2.1. Then*

$$u(x) \leq C_n \|f\|_{L^\infty(D)} \left(\operatorname{dist}(x, \Omega^c(u))\right)^2,$$

as long as $2\operatorname{dist}(x, \Omega^c(u)) < \operatorname{dist}(x, \partial D)$. □

In order to obtain $C^{1,1}$ estimates for the solutions of (2.1) we need to make a slightly stronger assumption on the function f. Namely, we require f to have a $C^{1,1}$ regular potential, i.e.

$$f = \Delta\psi \quad \text{in } D, \quad \text{with } \psi \in C^{1,1}(D). \tag{2.2}$$

We use the following second order derivative estimates associated with such f: if v is a solution of

$$\Delta v = f \quad \text{in} \quad B_{2R}(x^0) \subset D, \tag{2.3}$$

then

$$\|D^2 v\|_{L^\infty(B_R(x^0))} \leq C_n \left(\frac{\|v\|_{L^\infty(B_{2R}(x^0))}}{R^2} + \|D^2\psi\|_{L^\infty(B_{2R}(x^0))} \right). \tag{2.4}$$

We leave this estimate as an easy exercise to the reader.

Theorem 2.3 ($C^{1,1}$ regularity). *Let $u \in L^\infty(D)$, $u \geq 0$, satisfy* (2.1) *with f as in* (2.2). *Then $u \in C^{1,1}_{\mathrm{loc}}(D)$ and*

$$\|u\|_{C^{1,1}(K)} \leq C(\|u\|_{L^\infty(D)} + \|D^2\psi\|_{L^\infty(D)}),$$

for any $K \Subset D$, where $C = C(n, \operatorname{dist}(K, \partial D))$.

Proof. Since $u \in W^{2,p}_{\mathrm{loc}}$ and $D^2 u = 0$ a.e. on $\Omega^c(u)$, it will be enough to prove a uniform bound for $D^2 u$ in $\Omega(u) \cap K$ for $K \Subset D$.

To this end, fix $x^0 \in \Omega(u) \cap K$ and let $d = \operatorname{dist}(x^0, \Omega^c(u))$ and $\delta = \operatorname{dist}(K, \partial D)$. Then consider the following two possibilities.

1) $d < \delta/5$ (see Fig. 2.2). In this case, let $y_0 \in \partial B_d(x^0) \cap \partial\Omega$. Then $B_{4d}(y_0) \subset B_{5d}(x^0) \Subset D$. Applying Theorem 2.1, we have

$$\|u\|_{L^\infty(B_{2d}(y_0))} \leq C_n \|f\|_{L^\infty(D)}\, d^2.$$

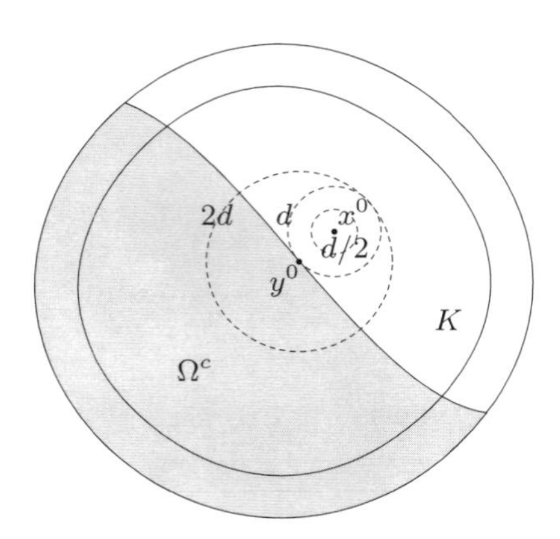

Figure 2.2. Case 1) $d < \delta/5$

Now note that $B_d(x^0) \subset B_{2d}(y_0)$ and $\Delta u = f$ in $B_d(x^0)$. By the interior estimate (2.4),

$$\|D^2 u\|_{L^\infty(B_{d/2}(x^0))} \le C_n(\|f\|_{L^\infty(D)} + \|D^2\psi\|_{L^\infty(D)}).$$

In fact, the term $\|f\|_{L^\infty(D)}$ is redundant as $\|f\|_{L^\infty(D)} \le C_n \|D^2\psi\|_{L^\infty(D)}$.

2) $d \ge \delta/5$. In this case, the interior derivative estimate for u in $B_d(x^0)$ gives

$$\|D^2 u\|_{L^\infty(B_{d/2}(x^0))} \le C_n \left(\frac{\|u\|_{L^\infty(D)}}{\delta^2} + \|D^2\psi\|_{L^\infty(D)} \right).$$

Combining cases 1) and 2), we obtain

$$\|u\|_{C^{1,1}(K)} \le C_n \left(\frac{\|u\|_{L^\infty(D)}}{\delta^2} + \|D^2\psi\|_{L^\infty(D)} \right). \qquad \square$$

2.2. ACF monotonicity formula and generalizations

The purpose of this section is to introduce an important technical tool, the so-called *monotonicity formula* of Alt-Caffarelli-Friedman [**ACF84**] (ACF for short) as well as some of its generalizations.

2.2.1. Harmonic functions. We start with a "baby" version of the ACF monotonicity formula for harmonic functions and show how to use it to obtain gradient estimates.

For a harmonic function u in the unit ball B_1 define the following quantity:

$$J(r,u) = \frac{1}{r^2} \int_{B_r} \frac{|\nabla u|^2 dx}{|x|^{n-2}}, \quad 0 < r < 1.$$

Then it is relatively straightforward to show that $r \mapsto J(r,u)$ is monotone nondecreasing. Indeed, if we represent u as a locally uniformly convergent series

$$u(x) = \sum_{k=0}^{\infty} f_k(x),$$

where $f_k(x)$ are homogeneous harmonic polynomials of degree k, and use the orthogonality of homogeneous harmonic polynomials of different degrees, we will have

$$\begin{aligned} J(r,u) &= \frac{1}{r^2}\int_0^r \int_{\partial B_1} |\nabla u(\rho\theta)|^2 \rho \, d\theta d\rho \\ &= \frac{1}{r^2}\int_0^r \int_{\partial B_1} \rho \sum_{k=1}^{\infty} |\nabla f_k(\rho\theta)|^2 d\theta d\rho \\ &= \frac{1}{r^2}\int_0^r \int_{\partial B_1} \sum_{k=1}^{\infty} \rho^{2k-1} |\nabla f_k(\theta)|^2 d\theta d\rho \\ &= \sum_{k=1}^{\infty} a_k r^{2(k-1)}, \end{aligned}$$

with

$$a_k = \frac{1}{2k}\int_{\partial B_1} |\nabla f_k(\theta)|^2 d\theta \geq 0.$$

This immediately implies that $r \mapsto J(r,u)$ is monotone nondecreasing.

We next illustrate how to use this monotonicity formula to obtain interior gradient estimates for harmonic functions.

a) Letting $r \to 0+$, we obtain

$$J(0+,u) \leq J(1/2,u).$$

On the other hand, since u is C^1 (actually real-analytic) near the origin, it is easy to see that $J(0+,u) = c_n |\nabla u(0)|^2$, for $c_n > 0$, which implies that

$$c_n |\nabla u(0)|^2 \leq J(1/2,u).$$

b) Next, we claim that $J(1/2,u)$ is controllable by the L^2-norm of u over B_1. More precisely, there exists a dimensional constant C_n such that

$$J(1/2,u) \leq C_n \|u\|^2_{L^2(B_1)}.$$

This can be viewed as a weighted form of the energy (or the Caccioppoli) inequality. To prove this statement, extend the kernel $|x|^{2-n}$ from $B_{1/2}$ to a function V on B_1 in a smooth nonnegative way, so that $V \equiv 0$ near ∂B_1. Also let $\delta > 0$ be a small number and $\hat{V} = \hat{V}_\delta = \min\{V, \delta^{2-n}\}$.

Then, using the equality $|\nabla u|^2 = \Delta(u^2/2)$, we have

$$\begin{aligned}\int_{B_{1/2}\setminus B_\delta} \frac{|\nabla u|^2}{|x|^{n-2}} dx &\le \int_{B_1} \Delta\left(\frac{u^2}{2}\right)\hat{V} dx\\ &= -\int_{\partial B_\delta} \frac{u^2}{2}(n-2)\delta^{1-n} dH^{n-1} + \int_{B_1\setminus B_\delta}\left(\frac{u^2}{2}\right)\Delta V\\ &\le \int_{B_1\setminus B_{1/2}}\left(\frac{u^2}{2}\right)\Delta V dx,\end{aligned}$$

which, upon letting $\delta \to 0$, gives

$$J(1/2, u) \le C_n \|u\|^2_{L^2(B_1)}.$$

Combining the estimates in a) and b) we arrive at

$$|\nabla u(0)| \le C_n \|u\|_{L^2(B_1)}.$$

Obviously, this is not the best way to establish the inequality above. This method is rather an illustration of the ability of monotonicity formulas to produce such gradient estimates.

2.2.2. ACF monotonicity formula.

Theorem 2.4 (Alt-Caffarelli-Friedman (ACF) monotonicity formula). *Let $u_\pm$ be a pair of continuous functions such that*

$$u_\pm \ge 0, \quad \Delta u_\pm \ge 0, \quad u_+ \cdot u_- = 0 \quad \text{in } B_1.$$

Then the functional

$$\begin{aligned} r \mapsto \Phi(r) = \Phi(r, u_+, u_-) &:= J(r, u_+) J(r, u_-)\\ &= \frac{1}{r^4} \int_{B_r} \frac{|\nabla u_+|^2 dx}{|x|^{n-2}} \int_{B_r} \frac{|\nabla u_-|^2 dx}{|x|^{n-2}}\end{aligned}$$

is nondecreasing for $0 < r < 1$.

Remark 2.5. Note that it follows from Exercise 2.4 that $J(r, u_\pm) < \infty$ for $r \in (0, 1)$. Sometimes, abusing the terminology, we will say that u_+ and u_- have disjoint "supports" instead of properly saying disjoint positivity sets $\{u_\pm > 0\}$.

Example 2.6. Each of the terms $J(r, u_\pm)$ can be understood as a weighted average of $|\nabla u_\pm|^2$. Thus, if $u_\pm = k_\pm x_1^\pm$, then

$$J(r, u_\pm) \equiv c_n k_\pm^2, \quad \Phi(r, u_+, u_-) \equiv c_n^2 k_+^2 k_-^2.$$

More generally, if $S_\pm = \partial\{u_\pm > 0\}$ are smooth hypersurfaces touching at the origin, and ν is their common normal there (see Fig. 2.3), then

$$J(0+, u_\pm) = c_n (\partial_\nu u_\pm(0))^2.$$

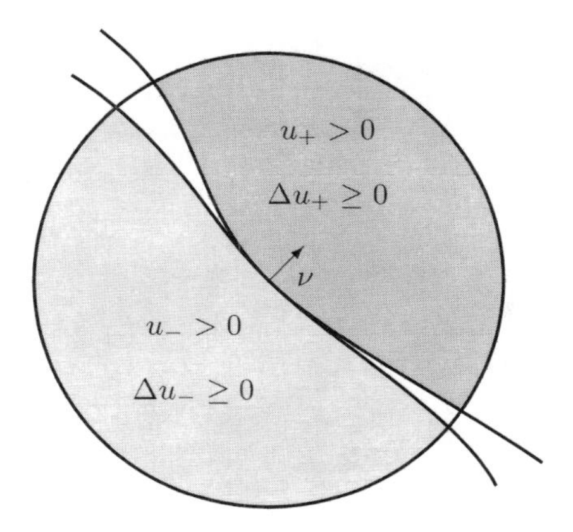

Figure 2.3. ACF monotonicity formula

In particular, the monotonicity formula implies that

$$c_n^2(\partial_\nu u_+(0))^2(\partial_\nu u_-(0))^2 \leq \Phi(1/2, u_+, u_-).$$

We next consider a more insightful example.

Example 2.7 (Friedland-Hayman inequality)**.** Let $\mathcal{C}$ be a cone with vertex at the origin generated by an open subset $\Sigma_0 \subset \partial B_1$, i.e.

$$\mathcal{C} = \{r\,\theta : r > 0,\ \theta \in \Sigma_0\}.$$

Consider a homogeneous harmonic function in $\mathcal{C}$ of the form

$$h(r\,\theta) = r^\alpha f(\theta), \qquad \alpha > 0,$$

vanishing on $\partial\mathcal{C}$. We have

$$\begin{aligned}\Delta h &= \partial_{rr} h + \frac{n-1}{r}\,\partial_r h + \frac{1}{r^2}\Delta_\theta h\\ &= r^{\alpha-2}[(\alpha(\alpha-1) + (n-1)\alpha) f(\theta) + \Delta_\theta f(\theta)],\end{aligned}$$

where Δ_θ is the spherical Laplacian (which is the same as the Laplace-Beltrami operator on the unit sphere). Thus, h is harmonic in $\mathcal{C}$ iff f is an eigenfunction of Δ_θ in Σ_0:

$$-\Delta_\theta f(\theta) = \lambda f(\theta) \quad \text{in } \Sigma_0,$$

where

$$\lambda = \alpha(n-2+\alpha).$$

If $h > 0$ in Σ_0, then λ above will be the principal eigenvalue of the spherical Laplacian in Σ_0 and the corresponding α will be denoted by $\alpha(\Sigma_0)$ and called the characteristic constant of Σ_0.

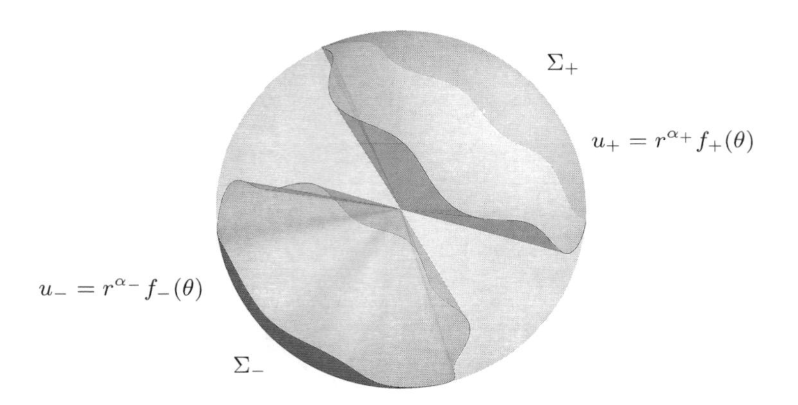

Figure 2.4. Two homogeneous harmonic functions $u_\pm$ in cones generated by the spherical regions $\Sigma_\pm$

Now, take two disjoint open sets $\Sigma_\pm$ on the unit sphere, find their principal eigenvalues $\lambda_\pm$ and the corresponding eigenfunctions $f_\pm$, and then form homogeneous harmonic functions

$$u_\pm = r^{\alpha_\pm} f_\pm(\theta) \quad \text{in } \mathcal{C}_\pm = \{r\,\theta : r > 0,\ \theta \in \Sigma_\pm\},$$

where $\alpha_\pm = \alpha(\Sigma_\pm) > 0$ are the characteristic constants of $\Sigma_\pm$ (see Fig. 2.4). If now we extend $u_\pm$ from $\mathcal{C}_\pm$ to the entire $\mathbb{R}^n$ by zero in the complements of $\mathcal{C}_\pm$, respectively, then the resulting functions will be subharmonic in $\mathbb{R}^n$ (see Exercise 2.6) and the pair u_+ and u_- will satisfy the assumptions of the ACF monotonicity formula. On the other hand, it is easy to compute that

$$\Phi(r, u_+, u_-) = C r^{2(\alpha_+ + \alpha_- - 2)}$$

for some $C = C(n, f_\pm) > 0$, and therefore the monotonicity formula in this particular case is equivalent to the inequality.

$$\alpha_+ + \alpha_- \geq 2.$$

This inequality has been first established by Friedland and Hayman in [**FH76**]. What is interesting is that it actually implies the monotonicity formula for all $u_\pm$ as in Theorem 2.4, not necessarily homogeneous, as we show below. We refer to the book of Caffarelli-Salsa [**CS05**, Chapter 12], for a detailed proof of the Friedland-Hayman inequality in any dimension $n \geq 2$. However, the case $n = 2$ is rather elementary and is left as an exercise to the reader (Exercise 2.7).

2.2.2.1. *Reduction of ACF monotonicity formula to Friedland-Hayman inequality.* Here we mostly follow Caffarelli [**Caf98b**].

We start with a remark that the functional J scales linearly, in the sense that if

$$u_\lambda(x) = \frac{1}{\lambda} u(\lambda x),$$

then

$$J(r/\lambda, u_\lambda) = J(r, u).$$

In particular, this implies that we can assume $u_\pm$ to be defined in B_R for a certain $R > 1$. Then it will suffice to show that $\Phi'(r) \geq 0$ only for $r = 1$.

It will be convenient to introduce

$$I(r, u) = \int_{B_r} \frac{|\nabla u|^2}{|x|^{n-2}} dx.$$

Thus, $J(r, u) = \frac{1}{r^2} I(r, u)$ and $\Phi(r, u_+, u_-) = \frac{1}{r^4} I(r, u_+) I(r, u_-)$. According to Exercise 2.4 we may assume that $I(1, u_\pm) < \infty$. Besides, since $I(r, u_\pm)$ are absolutely continuous functions of r, we may assume that $r = 1$ is a Lebesgue point for both of them. For the sake of brevity, we will denote $I_\pm = I(1, u_\pm)$ and $I'_\pm = I'(1, u_\pm)$. Then

$$\Phi'(1) = I'_+ I_- + I_+ I'_- - 4 I_+ I_-$$

and we want to show that

$$\frac{I'_+}{I_+} + \frac{I'_-}{I_-} \geq 4.$$

We now rewrite this as an inequality on the unit sphere. To further simplify the notation, we will use u to denote either u_+ or u_-. For the next estimate, let u_ε be a mollification of u, which still satisfies $\Delta u_\varepsilon \geq 0$, $u_\varepsilon \geq 0$. Using that $|\nabla u_\varepsilon|^2 \leq \Delta(u_\varepsilon^2/2)$ and that $\Delta(1/|x|^{n-2})$ is a nonpositive measure, we obtain

$$\begin{aligned} I(1, u_\varepsilon) &= \int_{B_1} \frac{|\nabla u_\varepsilon|^2}{|x|^{n-2}} dx \leq \int_{B_1} \frac{\Delta\left(\frac{u_\varepsilon^2}{2}\right)}{|x|^{n-2}} dx \\ &\leq \int_{\partial B_1} \left(u_\varepsilon \partial_r u_\varepsilon + \frac{n-2}{2} u_\varepsilon^2 \right) d\theta. \end{aligned}$$

Letting $\varepsilon \to 0+$ gives

$$I(1, u) \leq \int_\Sigma \left(u\, \partial_r u + \frac{n-2}{2} u^2 \right) d\theta,$$

where $\Sigma = \{u > 0\} \cap \partial B_1$. On the other hand, by Exercise 2.5,

$$I'(1, u) = \int_\Sigma |\nabla u|^2 d\theta.$$

Thus,

$$\frac{I'(1,u)}{I(1,u)} \geq \frac{\displaystyle\int_\Sigma [(\partial_r u)^2 + |\nabla_\theta u|^2] d\theta}{\displaystyle\int_\Sigma \left[u\,\partial_r u + \frac{n-2}{2}u^2\right] d\theta}.$$

Next note that

$$\frac{\displaystyle\int_\Sigma |\nabla_\theta u|^2}{\displaystyle\int_\Sigma u^2} \geq \lambda,$$

where $\lambda = \lambda(\Sigma)$ is the principal eigenvalue of the spherical Laplacian Δ_θ in Σ, so we want to split $u\,\partial_r u$ in an optimal fashion, to spread its control between $\int(\partial_r u)^2$ and $\int|\nabla_\theta u|^2$, i.e.,

$$\int_\Sigma u\,\partial_r u \leq \frac{1}{2}\left[\alpha\int_\Sigma u^2 + \frac{1}{\alpha}\int_\Sigma (\partial_r u)^2\right],$$

where the constant $\alpha > 0$ is to be chosen below. This will leave us with

$$\frac{I'(1,u)}{I(1,u)} \geq 2\frac{\displaystyle\int_\Sigma (\partial_r u)^2 + \lambda u^2}{\displaystyle\frac{1}{\alpha}\int_\Sigma (\partial_r u)^2 + (\alpha + n - 2)\int_\Sigma u^2}.$$

To perfectly balance both terms, we need

$$\frac{1}{\alpha} = \frac{\alpha + n - 2}{\lambda}, \quad \text{or} \quad \alpha(\alpha + n - 2) = \lambda;$$

in other words, $\alpha = \alpha(\Sigma)$ is the characteristic constant of Σ. This choice will give us

$$\frac{I'(1,u)}{I(1,u)} \geq 2\alpha.$$

Hence, if $\Sigma_\pm = \{u_\pm > 0\} \cap \partial B_1$ and $\alpha_\pm = \alpha(\Sigma_\pm)$, then

$$\frac{I'_+}{I_+} + \frac{I'_-}{I_-} - 4 \geq 2(\alpha_+ + \alpha_- - 2).$$

Since $\Sigma_\pm$ are disjoint open sets on ∂B_1, the required inequality will follow from the Friedland-Hayman inequality

$$\alpha_+ + \alpha_- - 2 \geq 0.$$

2.2.3. Generalizations. If u is a nonnegative subharmonic function, then $J(r, u)$ can be controlled in terms of L^2-norm of u, precisely as we have done for harmonic functions in §2.2.1. The only difference is that we have to use the inequality $|\nabla u|^2 \leq \Delta(u^2/2)$ in place of the equality there; see Exercises 2.3, 2.4. Thus, we still have that

$$J(1/2, u) \leq C_n \|u\|^2_{L^2(B_1)}.$$

Consequently, we obtain the following variant of the monotonicity formula, which takes the form of an estimate.

Theorem 2.8 (ACF estimate). *Let $u_\pm$ be as in Theorem 2.4. Then*

$$\Phi(r, u_+, u_-) \leq C_n \|u_+\|^2_{L^2(B_1)} \|u_-\|^2_{L^2(B_1)},$$

for $0 < r \leq 1/2$.

In some applications, this weaker form of the monotonicity formula turns out to be sufficient. However, in other applications, one needs to use Theorem 2.4 at its full strength; moreover, one needs to have information on the case of $\Phi(r)$ being a constant in some interval.

Theorem 2.9 (Case of equality in ACF monotonicity formula). *Let $u_\pm$ be as in Theorem 2.4 and suppose that $\Phi(r_1) = \Phi(r_2)$ for some $0 < r_1 < r_2 < 1$. Then either one or the other of the following holds:*

(i) *$u_+ = 0$ in B_{r_2} or $u_- = 0$ in B_{r_2};*

(ii) *there exist a unit vector e and constants $k_\pm > 0$ such that*

$$u_+(x) = k_+(x \cdot e)^+, \quad u_-(x) = k_-(x \cdot e)^- \quad \text{in } B_{r_2}.$$

The proof is based on the following lemma.

Lemma 2.10. *Let $u_\pm$ be as in Theorem 2.4 and suppose that $\Phi(r_1) = \Phi(r_2) > 0$ for some $0 < r_1 < r_2 < 1$. Then*

(1) *$u_+ \Delta u_+ = u_- \Delta u_- = 0$ on B_{r_2} in the sense of measures;*

(2) *for every $r_1 < r < r_2$, $\{u_\pm > 0\} \cap \partial B_r$ are complementary half-spherical caps and $u_\pm$ are principal eigenfunctions of the spherical Laplacian on $\{u_\pm > 0\} \cap \partial B_r$.*

Proof. Analyzing the reduction of the ACF monotonicity formula to the Friedland-Hayman inequality in §2.2.2.1, one easily sees (1) and the part of (2) that $u_\pm$ must be eigenfunctions $\{u_\pm > 0\} \cap \partial B_r$ for the spherical Laplacian. The fact that $\{u_\pm > 0\} \cap \partial B_r$ are half-spherical caps follows from the case of equality in the Friedland-Hayman inequality: $\alpha_+ + \alpha_- = 2$ if and only if $\Sigma_\pm$ are half-spheres. To see the latter, we refer to Caffarelli-Karp-Shahgholian [**CKS00**, Lemma 2.3], where it is shown that the spherical caps $\Sigma^*_\pm$ of the same spherical measure as $\Sigma_\pm$ must be half-spheres. On the other hand, since the principal eigenfunctions of the spherical Laplacian are real-analytic and positive, the result of Kawohl [**Kaw86**, Theorem 2.1 and Remark 2.6] implies that $\alpha(\Sigma^*) < \alpha(\Sigma)$ unless Σ is a spherical cap. □

Proof of Theorem 2.9. If $\Phi(r_1) = \Phi(r_2) = 0$, then it is easy to see that alternative (i) must hold. Thus, in what follows we may assume that $\Phi(r_1) = \Phi(r_2) > 0$.

To proceed, we observe that the principal eigenfunctions of the spherical Laplacian on the half-spherical caps are linear functions. Then, if u stands for any of the functions $u_\pm$, from Lemma 2.10 we immediately obtain a representation

$$u(x) = [a(r) \cdot x]^+, \quad \text{for } r_1 < r = |x| < r_2,$$

where $a(r)$ is an $\mathbb{R}^n$-valued function on (r_1, r_2). Moreover, $a(r) \neq 0$ for any $r_1 < r < r_2$, otherwise we will have $\Phi(r) = \Phi(r_1) = 0$.

Next, from Lemma 2.10 we also have that u is harmonic in $\{u > 0\} \cap B_{r_2}$. Thus, $a(r)$ is real-analytic in (r_1, r_2) and we can explicitly compute Δu to obtain

$$\left[a''(r) + \frac{n+1}{r} a'(r)\right] \cdot x = 0, \quad \text{for } r_1 < r = |x| < r_2.$$

Consequently,

$$a''(r) + \frac{n+1}{r} a'(r) = 0, \quad r_1 < r < r_2,$$

and

$$a(r) = \frac{c}{r^n} + b, \quad r_1 < r < r_2,$$

for some constant vectors c and b. We thus obtain representations

$$u_\pm(x) = \left[\left(\frac{c_\pm}{|x|^n} + b_\pm\right) \cdot x\right]^+, \quad r_1 < |x| < r_2.$$

Because of the uniqueness of harmonic continuation, the representations above will continue to hold for $\rho < |x| < r_2$, as long as $a_\pm(r) = c_\pm r^{-n} + b_\pm$ do not vanish in (ρ, r_2). Moreover, from real analyticity arguments, $\Phi(r)$ must also be constant on (ρ, r_2). In particular, $\Phi(\rho) = \Phi(r_2) \neq 0$ and therefore $a_\pm(\rho) \neq 0$. Taking the infimum ρ_* among all ρ as above, we must have $\rho_* = 0$ and consequently

$$u_\pm(x) = \left[\left(\frac{c_\pm}{|x|^n} + b_\pm\right) \cdot x\right]^+, \quad 0 < |x| < r_2.$$

Since $u_\pm$ are continuous at the origin, this is possible only if $c_\pm = 0$, and therefore

$$u_\pm(x) = (b_\pm \cdot x)^+, \quad |x| < r_2.$$

Finally, it is clear that $b_\pm$ must point in opposite directions, which implies alternative (ii) in the statement of Theorem 2.9. □

Next, we state a generalization of the ACF monotonicity formula, due to Caffarelli-Jerison-Kenig [**CJK02**].

Theorem 2.11 (Caffarelli-Jerison-Kenig (CJK) estimate). *Let $u_\pm$ be a pair of continuous functions in B_1 satisfying*

$$u_\pm \geq 0, \quad \Delta u_\pm \geq -1, \quad u_+ \cdot u_- = 0 \quad \text{in } B_1.$$

Then

$$\Phi(r, u_+, u_-) \leq C_n(1 + J(1, u_+) + J(1, u_-))^2, \quad 0 < r < 1.$$

This estimate still has some features of the ACF monotonicity formula, so sometimes it is referred to as the CJK *almost monotonicity formula.* The proof utilizes careful iterative estimates based on the energy and Friedland-Hayman inequalities. We refer to the original paper [**CJK02**] for the proof.

We will also need a scaled version of Theorem 2.11.

Theorem 2.11′. *Let $u_\pm$ be a pair of continuous functions in B_R satisfying*

$$u_\pm \geq 0, \quad \Delta u_\pm \geq -L, \quad u_+ \cdot u_- = 0 \quad \text{in } B_R.$$

Then

$$\Phi(r, u_+, u_-) \leq C_n(R^2L^2 + J(R, u_+) + J(R, u_-))^2, \quad 0 < r < R.$$

If u is a nonnegative continuous function such that $\Delta u \geq -1$ in B_1, then using $2|\nabla u|^2 \leq \Delta(u^2) + 2u$, one can show that

$$J(1/2, u) \leq C_n \left(1 + \|u\|^2_{L^2(B_1)}\right).$$

This leads to the following form of the CJK estimate, akin to Theorem 2.8.

Theorem 2.12. *Let $u_\pm$ be as in Theorem* 2.11. *Then*

$$\Phi(r, u_+, u_-) \leq C_n \left(1 + \|u_+\|^2_{L^2(B_1)} + \|u_-\|^2_{L^2(B_1)}\right)^2,$$

for $0 < r \leq 1/2$.

In fact, later in this chapter we will need a scaled version of this estimate.

Theorem 2.12′. *Let $u_\pm$ be as in Theorem* 2.11′. *Then*

$$\Phi(r, u_+, u_-) \leq C_n \left(R^2L^2 + \frac{\|u_+\|^2_{L^2(B_R)} + \|u_-\|^2_{L^2(B_R)}}{R^{n+2}}\right)^2,$$

for $0 < r \leq R/2$.

A stronger statement can be proved about $\Phi(r)$ that retains more features of the monotonicity formula at the expense of an additional assumption on the growth of functions $u_\pm$ near the origin.

Theorem 2.13. *Let $u_\pm$ be as in Theorem 2.11 and assume additionally that $u_\pm(x) \le C_0|x|^\varepsilon$ in B_1 for some $\varepsilon > 0$. Then there exists $C_1 = C(C_0, n, \varepsilon)$ such that*

$$\Phi(r_1) \le (1 + r_2^\varepsilon)\Phi(r_2) + C_1 r_2^{2\varepsilon},$$

for $0 < r_1 \le r_2 < 1$. In particular, the limit $\Phi(0+)$ exists.

For the proof, we again refer to [**CJK02**].

In this book, every time we apply the CJK estimate, the functions $u_\pm$ will satisfy the growth condition above, and therefore we may use this stronger statements about Φ.

2.3. Optimal regularity in obstacle-type problems

In this section, following the ideas in [**Sha03**], we use the CJK estimate (Theorem 2.12) to prove the optimal regularity for the functions u solving an obstacle type problem $\mathbf{OT}_1$–$\mathbf{OT}_2$.

Theorem 2.14 ($C^{1,1}$ regularity). *Let $u \in L^\infty(D)$ satisfy $\mathbf{OT}_1$–$\mathbf{OT}_2$. Then $u \in C^{1,1}_{\mathrm{loc}}(D)$ and*

$$\|u\|_{C^{1,1}(K)} \le CM\left(1 + \|u\|_{L^\infty(D)} + \|g\|_{L^\infty(D)}\right),$$

for any $K \Subset D$, where $C = C(n, \operatorname{dist}(K, \partial D))$ and $M = \max\{1, M_1, M_2\}$.

One of the important ingredients in the proof is the following fundamental lemma. Basically, it says that the positive and negative parts of the directional derivatives of solutions to obstacle-type problems satisfy the assumptions in the CJK estimate.

Lemma 2.15. *Let $u \in C^1(D)$ satisfy $\mathbf{OT}_1$–$\mathbf{OT}_2$. Then for any unit vector e,*

$$\Delta(\partial_e u)^\pm \ge -L \quad \text{in } D,$$

where $L = M_1 + M_2\|\nabla u\|_{L^\infty(D)}$.

Proof. Fix a direction e and let $v = \partial_e u$. Also let

$$E := \{v > 0\}.$$

Note that $E \subset G$ because of the second condition in $\mathbf{OT}_1$. Then, formally, for $x \in E$,

$$\begin{aligned}\Delta(v^+) = \partial_e \Delta u(x) &= e \cdot \nabla_x f(x, u) + \partial_u f(x, u) D_e u\\ &\ge -M_1 - M_2\|\nabla u\|_{L^\infty(D)} =: -L.\end{aligned}$$

To justify this computation, observe that $\Delta(v^+) \geq -L$ in D is equivalent to the inequality

$$-\int_D \nabla(v^+)\nabla\eta\,dx \geq -L\int_D \eta\,dx, \tag{2.5}$$

for any nonnegative $\eta \in C_0^\infty(D)$. Suppose first that $\operatorname{supp}\eta \subset \{v > \delta\}$ with $\delta > 0$. Then writing the equation

$$-\int_D \nabla u\nabla\eta\,dx = \int_D f\eta\,dx$$

with $\eta = \eta(x)$ and $\eta = \eta(x - he)$, we obtain an equation for the incremental quotient

$$v_{(h)}(x) := \frac{u(x+he) - u(x)}{h}.$$

Namely, we obtain

$$-\int_D \nabla v_{(h)}\nabla\eta\,dx = \frac{1}{h}\int_D [f(x+he, u(x+he)) - f(x, u(x))]\eta\,dx, \tag{2.6}$$

for small $h > 0$. Note that $u(x+he) > u(x)$ on $\operatorname{supp}\eta \subset \{v > \delta\}$, and from the hypotheses on f we have

$$\begin{aligned} f(x+he, u(x+he)) &- f(x, u(x)) \\ &= [f(x+he, u(x+he)) - f(x+he, u(x))] \\ &\quad + [f(x+he, u(x)) - f(x, u(x))] \\ &\geq -M_1 h - M_2[u(x+he) - u(x)], \end{aligned}$$

for small h. Letting in (2.6) $h \to 0$ and then $\delta \to 0$ we arrive at

$$\begin{aligned} -\int_D \nabla v\nabla\eta\,dx &\geq -\int_D (M_1 + M_2 v)\eta dx \\ &\geq -L\int_D \eta\,dx, \end{aligned}$$

for arbitrary $\eta \geq 0$ with $\operatorname{supp}\eta \Subset \{v > 0\}$.

Thus, we have proved that $\Delta v \geq -L$ in the open set $E = \{v > 0\}$ in the sense of distributions. Then it is a simple exercise to show that (2.5) holds for any nonnegative $\eta \in C_0^\infty(D)$; see Exercise 2.6.

To prove the same inequality for v^-, we simply reverse the direction of the vector e. □

Remark 2.16. In the particular case of Problems **A**, **B**, **C**, the conditions **OT**$_2$ are satisfied with $M_1 = M_2 = 0$, which implies that one can take $L = 0$ in Lemma 2.15. In particular, for a solution u of Problem **A**, **B**, or **C**, $(\partial_e u)^\pm$ are subharmonic functions.

We are now ready to prove Theorem 2.14.

Proof of Theorem 2.14. We start by observation that at any Lebesgue point x^0 of D^2u the function u is twice differentiable, since $u \in W^{2,p}_{\mathrm{loc}}(D)$ with $p > n$; see e.g. [**Eva98**, Theorem 5.8.5]. Then fix such a point $x^0 \in K \Subset D$ where u is twice differentiable and define

$$v(x) = \partial_e u(x),$$

for a unit vector e orthogonal to $\nabla u(x^0)$ (if $\nabla u(x^0) = 0$, take an arbitrary unit e). Without loss of generality we may assume that $x^0 = 0$. Our aim is to obtain a uniform estimate for $\partial_{x_j e} u(0) = \partial_{x_j} v(0)$, $j = 1, \dots, n$. By construction, $v(0) = 0$ and v is differentiable at 0. Hence, we have the Taylor expansion

$$v(x) = \xi \cdot x + o(|x|), \quad \xi = \nabla v(0).$$

Now, if $\xi = 0$ then $\partial_{x_j} v(0) = 0$ for all $j = 1, \dots, n$ and there is nothing to estimate. If $\xi \neq 0$, consider the cone

$$\mathcal{C} = \{x \in \mathbb{R}^n : \xi \cdot x \geq |\xi||x|/2\},$$

which has the property that

$$\mathcal{C} \cap B_r \subset \{v > 0\}, \quad -\mathcal{C} \cap B_r \subset \{v < 0\},$$

for sufficiently small $r > 0$. Consider also the rescalings

$$v_r(x) = \frac{v(rx)}{r}, \quad x \in B_1.$$

Note that $v_r(x) \to v_0(x) := \xi \cdot x$ uniformly in B_1 and $\nabla v_r \to \nabla v_0$ in $L^p(B_1)$, $p > n$. The latter follows from the equality

$$\int_{B_1} |\nabla v_r(x) - \xi|^p dx = \frac{1}{r^n} \int_{B_r} |\nabla v(x) - \nabla v(0)|^p dx,$$

where the right-hand side goes to zero as $r \to 0$, since $x^0 = 0$ is a Lebesgue point for ∇v. Then, for a positive dimensional constant c_n, we have

$$\begin{aligned}
c_n^2 |\xi|^4 &= \int_{\mathcal{C} \cap B_1} \frac{|\nabla v_0(x)|^2 dx}{|x|^{n-2}} \int_{-\mathcal{C} \cap B_1} \frac{|\nabla v_0(x)|^2 dx}{|x|^{n-2}} \\
&= \lim_{r \to 0} \int_{\mathcal{C} \cap B_1} \frac{|\nabla v_r(x)|^2 dx}{|x|^{n-2}} \int_{-\mathcal{C} \cap B_1} \frac{|\nabla v_r(x)|^2 dx}{|x|^{n-2}} \\
&= \lim_{r \to 0} \frac{1}{r^4} \int_{\mathcal{C} \cap B_r} \frac{|\nabla v(x)|^2 dx}{|x|^{n-2}} \int_{-\mathcal{C} \cap B_r} \frac{|\nabla v(x)|^2 dx}{|x|^{n-2}} \\
&\leq \lim_{r \to 0} \Phi(r, v^+, v^-),
\end{aligned}$$

where Φ is as in the ACF monotonicity formula (Theorem 2.4). In the next step we apply the CJK estimate (Theorem 2.12); however, we should suitably adjust (scale) $v^{\pm}$ first. Let $\delta = \frac{1}{2}\operatorname{dist}(K, \partial D)$ and $K_\delta = \{x : \operatorname{dist}(x, K) < \delta\}$. By Lemma 2.15, we have $\Delta v^{\pm} \geq -L_\delta$ in K_δ, where

$$L_\delta = M\left(1 + \|\nabla u\|_{L^\infty(K_\delta)}\right)$$

with $M = \max\{1, M_1, M_2\}$. Now, applying the scaled form of the CJK estimate in Theorem 2.12′, we obtain

$$c_n^2|\xi|^4 \leq \liminf_{r\to 0} \Phi(r, v^+, v^-) \leq C\left(L_\delta^2\delta^2 + \frac{\|\nabla u\|^2_{L^\infty(K_\delta)}}{\delta^2}\right)^2 \leq C(n,\delta)L_\delta^4.$$

Hence, we obtain that $|\xi| \leq C(n,\delta)L_\delta$. Now, using that

$$L_\delta = M(1 + \|\nabla u\|_{L^\infty(K_\delta)}) \leq C(n,\delta)N,$$

where

$$N = M(1 + \|u\|_{L^\infty(D)} + \|g\|_{L^\infty(D)}),$$

and recalling that $\xi = \nabla\partial_e u(x^0)$, we arrive at

$$|\nabla\partial_e u(x^0)| \leq C(n,\delta)N.$$

This does not give the desired estimate on $|D^2u|$ yet, since e is subject to the condition $e \cdot \nabla u(x^0) = 0$, unless $\nabla u(x^0) = 0$. If $\nabla u(x^0) \neq 0$, we may choose the coordinate system so that $\nabla u(x^0)$ be parallel to e_n. Then, taking $e = e_1, \dots, e_{n-1}$ in the estimate above, we obtain

$$|\partial_{x_i x_j} u(x^0)| \leq C(n,\delta)N, \quad i = 1, \dots, n-1, \quad j = 1, 2, \dots, n.$$

To obtain the estimate in the missing direction e_n, we use the equation $\Delta u = g$:

$$\begin{aligned}|\partial_{x_n x_n} u(x^0)| &\leq |\Delta u(x^0)| + |\partial_{x_1 x_1} u(x^0)| + \dots + |\partial_{x_{n-1}x_{n-1}} u(x^0)| \\ &\leq \|g\|_{L^\infty(D)} + C(n,\delta)N \leq C(n,\delta)N.\end{aligned}$$

This completes the proof of the theorem. □

2.4. Optimal regularity up to the fixed boundary

In this section we show that the $C^{1,1}$ estimate can be extended up to the boundary of the domain D. This will be particularly important in Chapter 8, where we study the behavior of the free boundary near ∂D. Because of the technical difficulty of the problem, we will restrict our study to the case of flat boundaries and zero Dirichlet data.

Theorem 2.17. *Let $u \in L^\infty_{\mathrm{loc}}(B_1^+)$ solve* $\mathbf{OT}_1$–$\mathbf{OT}_2$ *in $D = B_1^+$. Assume also that*

$$u = 0 \quad \text{on } B_1', \tag{2.7}$$

in the Sobolev trace sense. Then $u \in C^{1,1}(B^+_{1/4} \cup B'_{1/4})$ *and*

$$\|D^2 u\|_{L^\infty(B^+_{1/4})} \leq C_n M \left(1 + \|u\|_{L^\infty(B^+_1)} + \|g\|_{L^\infty(B^+_1)}\right),$$

where $M = \max\{1, M_1, M_2\}$.

Note that the condition (2.7) should be understood in the sense that $u\zeta \in W^{1,2}_0(B^+_1)$ for any cutoff function $\zeta \in C^\infty_0(B_1)$.

The idea of the proof is still based on the application of the CJK estimate, similarly to the interior case (Theorem 2.14). However, we need additional estimates to have uniformity up to the boundary.

Lemma 2.18. *Let* u *be as in Theorem* 2.17 *and let* τ *be a direction tangential to* B'_1. *Then*

$$|\partial_\tau u(x)| \leq C_0 x_n, \quad x \in B^+_{1/2},$$

for $C_0 = C_n M(1 + \|u\|_{L^\infty(B_1+)} + \|g\|_{L^\infty(B^+_1)})$.

Proof. We first note that by the Calderón-Zygmund estimates, $u \in W^{2,p}(B^+_{3/4}) \cap C^{1,\alpha}(\overline{B^+_{3/4}})$ for any $1 < p < \infty$ and $0 < \alpha < 1$. This can be seen e.g. by extending u by odd reflection into B^-_1 and then noticing that the extended function has a bounded Laplacian.

Next, by Lemma 2.15,

$$\Delta(\partial_\tau u)^\pm \geq -L \quad \text{in } B^+_{3/4},$$

for $L \leq C_n M(\|u\|_{L^\infty} + \|g\|_{L^\infty})$. Now let h be the solution of the Dirichlet problem

$$\Delta h = -L \quad \text{in } B^+_{3/4}, \qquad h = |\partial_\tau u| \quad \text{on } \partial B^+_{3/4}.$$

Then by the comparison principle

$$(\partial_\tau u)^\pm \leq h \quad \text{on } B^+_{3/4}.$$

Applying up-to-the-boundary elliptic L^p estimates, we obtain

$$\|\nabla h\|_{L^\infty(B^+_{1/2})} \leq C_0 = C_n M(\|u\|_{L^\infty} + \|g\|_{L^\infty}),$$

which along with the equality $h = \partial_\tau u = 0$ on $\partial B^+_{1/2} \cap B_{1/2}$ gives

$$h \leq C_0 x_n \quad \text{in } B^+_{1/2}.$$

Hence,

$$|\partial_\tau u| \leq C_0 x_n \quad \text{in } B^+_{1/2}$$

and the proof is complete. □

Lemma 2.19. *Let u be as in Theorem 2.17, $x^0 \in B^+_{1/4}$, and let $r = \frac{1}{2}\operatorname{dist}(x^0, B'_1) = \frac{1}{2}x^0_n$. Then*

$$\frac{1}{r^2}\int_{B_r(x^0)} \frac{|D^2u|^2}{|x-x^0|^{n-2}}\,dx \le C, \tag{2.8}$$

where $C = C_n M^2\left(1 + \|u\|^2_{L^\infty} + \|g\|^2_{L^\infty}\right)$.

Proof. Observe that it is enough to show that

$$J(r, \partial_\tau u) = \frac{1}{r^2}\int_{B_r(x^0)} \frac{|\nabla(\partial_\tau u)|^2}{|x-x^0|^{n-2}}\,dx \le C, \tag{2.9}$$

for any direction τ tangential to B'_1, since

$$|\partial_{x_n x_n} u| \le |g| + |\partial_{x_1 x_1} u| + \cdots + |\partial_{x_{n-1}x_{n-1}} u|$$

by the equation $\Delta u = g$ and consequently

$$|D^2u|^2 \le C_n\left(\|g\|^2_{L^\infty} + |\nabla(\partial_{x_1}u)|^2 + \cdots + |\nabla(\partial_{x_{n-1}}u)|^2\right).$$

In order to show (2.9), fix a tangential direction τ. Then the derivative $v = \partial_\tau u$ satisfies

$$\int_{B^+_1} \nabla v \nabla \eta\,dx = \int_{B^+_1} g\partial_\tau \eta\,dx, \tag{2.10}$$

for any $\eta \in W^{1,2}_0(B^+_1)$. Choose a test function

$$\eta = v\,\hat{G}\,\zeta^2,$$

where $\hat{G}$ is a truncation of the fundamental solution at x^0,

$$\hat{G}(x) = \min\{c_n(|x-x^0|^{2-n}, c_n\delta^{2-n}\},$$

for some small $\delta > 0$, and $\zeta \in C^\infty_0(B_{2r}(x^0))$ is a cutoff function such that

$$0 \le \zeta \le 1, \quad |\nabla\zeta| \le \frac{C_n}{r}, \quad |D^2\zeta| \le \frac{C_n}{r^2}$$

and $\zeta = 1$ on $B_r(x^0)$. Plugging η into (2.10), we obtain

$$\int \nabla v[(\nabla v)\hat{G}\zeta^2 + v\nabla(\hat{G}\zeta^2)]dx = \int g[(\partial_\tau v)\hat{G}\zeta^2 + v\partial_\tau(\hat{G}\zeta^2)]dx$$

and therefore

$$\begin{aligned}\int |\nabla v|^2\hat{G}\zeta^2 dx &= -\int \nabla\left(\tfrac{1}{2}v^2\right)\nabla(\hat{G}\zeta^2)dx \\ &\quad + \int g(\partial_\tau v)\hat{G}\zeta^2 dx + \int gv\partial_\tau(\hat{G}\zeta^2)dx \\ &= I_1 + I_2 + I_3.\end{aligned}$$

To estimate the integrals, note that by Lemma 2.18, $|v| \leq 4C_0 r$ in $B_{2r}(x^0)$. Then, integration by parts gives

$$\begin{aligned}
I_1 &= \int_{B_{2r(x^0)}\setminus B_\delta(x^0)} \tfrac{1}{2}v^2\Delta(G\zeta^2)dx - \int_{\partial B_\delta(x^0)} \tfrac{1}{2}v^2\partial_\nu G dH^{n-1}\\
&\leq \int_{B_{2r(x^0)}\setminus B_r(x^0)} \tfrac{1}{2}v^2[G\Delta(\zeta^2) + 2\nabla G\nabla(\zeta^2)]dx\\
&\leq C_n\|v\|^2_{L^\infty(B_{2r}(x^0))} \leq C_0^2 r^2.
\end{aligned}$$

Further,

$$I_3 \leq C_n r\|g\|_{L^\infty}\|v\|_{L^\infty(B_{2r}(x^0))} \leq C_0^2 r^2.$$

Finally, using that $|g\partial_T v| \leq \frac{1}{2}(|g|^2 + |\nabla v|^2)$, we obtain

$$I_2 \leq C_n r^2\|g\|^2_{L^\infty} + \frac{1}{2}\int |\nabla v|^2\hat{G}\zeta^2 dx \leq C_0^2 r^2 + \frac{1}{2}\int |\nabla v|^2\hat{G}\zeta^2 dx.$$

Thus, collecting all inequalities, we arrive at

$$\int |\nabla v|^2\hat{G}\zeta^2 dx \leq C_0^2 r^2.$$

Letting $\delta \to 0$, we obtain (2.9) and consequently (2.8). □

We are now ready to prove the $C^{1,1}$ regularity up to the boundary.

Proof of Theorem 2.17. Arguing as in the proof of Theorem 2.14, let $x^0 \in B^+_{1/4}$ be a Lebesgue point for D^2u. Also let e be a direction such that $\partial_e u(x^0) = 0$. Then if $v = \partial_e u$ and $\xi = \nabla v(x^0)$, we will have an estimate

$$c_n^2|\xi|^4 \leq \lim_{\rho\to 0}\Phi(\rho, v^+, v^-).$$

Now, keeping in mind Lemma 2.19, which says that

$$J(\delta, v) \leq C_n M^2\left(1 + \|u\|^2_{L^\infty(B_1^+)} + \|g\|^2_{L^\infty(B_1^+)}\right)$$

for $\delta = \frac{1}{2}\operatorname{dist}(x^0, B_1') = \frac{1}{2}x_n^0$, we want to apply the CJK estimate in Theorem 2.11′. Note that by Lemma 2.15,

$$\Delta v^\pm \geq -L = -C_n M\left(1 + \|u\|_{L^\infty(B_1^+)} + \|g\|_{L^\infty(B_1^+)}\right) \quad \text{in } B^+_{1/4}.$$

Then by Theorem 2.11′ we have

$$\begin{aligned}
c_n^2|\xi|^4 &\leq \lim_{\rho\to 0}\Phi(\rho, v^+, v^-) \leq C_n(L^2\delta^2 + J(\delta, v^+) + J(\delta, v^-))^2\\
&\leq C_n\left(L^2\delta^2 + J(\delta, v)\right)^2\\
&\leq C_n M^4\left(1 + \|u\|^2_{L^\infty(B_1^+)} + \|g\|^2_{L^\infty(B_1^+)}\right)^2,
\end{aligned}$$

independently of the distance of x^0 to B_1'. Finally, one can complete the proof exactly as in Theorem 2.14. □

2.5. A counterexample

In this section we describe a counterexample of Andersson-Weiss [**AW06**] of a non-$C^{1,1}$ two-dimensional solution of the equation

$$\Delta u = -\chi_{\{u>0\}},$$

known as the unstable obstacle problem. In particular, this shows the importance of the second condition in $\mathbf{OT}_2$ in Theorem 2.14.

Let B_1 be a unit ball in $\mathbb{R}^2$, $0 < \alpha < 1$, and m a positive integer. Consider then the subspace $C^\alpha_{*m}(\overline{B}_1)$ of $C^\alpha(\overline{B}_1)$ of functions $u(x_1, x_2)$ that satisfy

$$u(x_1, -x_2) = u(x_1, x_2), \qquad u \circ U_{2\pi/m} = u,$$

where U_θ is a rotation by an angle θ in the counterclockwise direction. In other words, functions in $C^\alpha_{*m}(\overline{B}_1)$ are obtained from their restrictions to the sector with $0 \leq \theta \leq \pi/m$ (using the polar coordinates) by even reflection with respect to the rays $\theta = k\pi/m$, $k = 0, \dots, 2m-1$.

Similarly, we define the subspace $C^\alpha_{*m}(\partial B_1)$.

Proposition 2.20. *Let $g \in C^\beta_{*m}(\partial B_1)$ for some $0 < \beta < 1$. Then there exists a constant $\kappa = \kappa(g)$ such that the boundary value problem*

$$\begin{aligned} \Delta u &= -\chi_{\{u>0\}} && \text{in} \quad B_1, \\ u &= g - \kappa && \text{on} \quad \partial B_1 \end{aligned}$$

has a solution $u \in C^\alpha_{*m}(B_1) \cap W^{2,p}_{\mathrm{loc}}(B_1)$, for some $\alpha = \alpha(g) > 0$ and all $1 < p < \infty$, which also satisfies $u(0) = |\nabla u(0)| = 0$.

Proof. For $\varepsilon > 0$ let $f_\varepsilon \in C^\infty(\mathbb{R})$ be a mollification of $\chi_{\{s>0\}}$ such that

$$\chi_{\{s>0\}} \leq f_\varepsilon(s) \leq \chi_{\{s>-\varepsilon\}}.$$

Consider now the operator $T_\varepsilon : C^0_{*m}(\overline{B}_1) \to C^0_{*m}(\overline{B}_1)$ for which $v = T_\varepsilon(u)$ is the solution of the Poisson problem

$$\begin{aligned} \Delta v &= -f_\varepsilon(u - u(0)) && \text{in} \quad B_1, \\ v &= g && \text{on} \quad \partial B_1. \end{aligned}$$

By the theory of strong solutions and the symmetry we have $T_\varepsilon(u) \in C^{\alpha_0}_{*m}(\overline{B}_1)$, for a constant $\alpha_0 = \alpha_0(g) > 0$ (see [**GT01**, Corollary 9.29]). Moreover,

$$\|T_\varepsilon u\|_{C^{\alpha_0}(\overline{B}_1)} \leq C(n, g),$$

for any $u \in C^0_{*m}(\overline{B}_1)$. As a consequence, T_ε is a compact operator on $C^\alpha_{*m}(\overline{B}_1)$ into itself for any $0 < \alpha < \alpha_0$. Then by the Schauder fixed point theorem there exists a fixed point u_ε of the operator T_ε,

$$T_\varepsilon(u_\varepsilon) = u_\varepsilon,$$

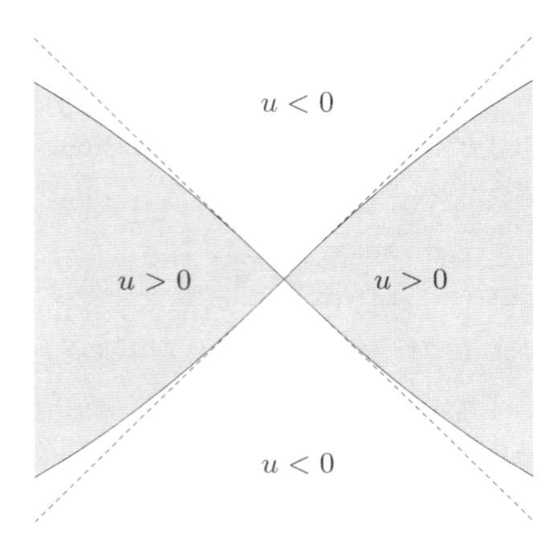

Figure 2.5. Non-$C^{1,1}$ solution of $\Delta u = -\chi_{\{u>0\}}$ with a cross-shaped singularity

or in other words, a solution of the semilinear problem

$$\begin{aligned} \Delta u_\varepsilon &= -f_\varepsilon(u_\varepsilon - u_\varepsilon(0)) \quad \text{in} \quad B_1, \\ u_\varepsilon &= g \qquad\qquad\qquad\quad\ \ \text{on} \quad \partial B_1. \end{aligned}$$

The family $\{u_\varepsilon\}$ is uniformly bounded in the C^α-norm on $\overline{B}_1$ and in the $W^{2,p}$-norm on any $K \Subset B_1$. Therefore, for a subsequence $\varepsilon = \varepsilon_k \to 0$, $u_\varepsilon - u_\varepsilon(0)$ converges uniformly to a function $u_0 \in C^\alpha(\overline{B}_1) \cap W^{2,p}_{\mathrm{loc}}(B_1)$. Moreover, since

$$\chi_{\{u_0>\delta\}} \leq f_\varepsilon(u_\varepsilon - u_\varepsilon(0)) \leq \chi_{\{u_0>-\delta\}}$$

for $\delta > 0$ and sufficiently small $\varepsilon > 0$, in the limit we obtain that

$$-\chi_{\{u_0\geq 0\}} \leq \Delta u_0 \leq -\chi_{\{u_0>0\}},$$

weakly in B_1. On the other hand, $\Delta u_0 = 0$ a.e. on $\{u_0 = 0\}$, since $u_0 \in W^{2,p}_{\mathrm{loc}}(B_1)$. Hence,

$$\Delta u_0 = -\chi_{\{u_0>0\}},$$

weakly in B_1. Moreover, it is immediate that $u_0 = g - \kappa$ on ∂B_1 for a constant $\kappa = \lim_{k\to\infty} u_{\varepsilon_k}(0)$. Finally, the equality $|\nabla u_0(0)| = 0$ follows from the symmetry, and the $C^{1,\alpha}_{\mathrm{loc}} \cap W^{2,p}_{\mathrm{loc}}$ regularity of u_0 follows from the elliptic L^p estimates (see Theorem 1.1). □

Proposition 2.21. *Let $u \in C^\alpha_{*2}(\overline{B_1}) \cap W^{2,p}_{\mathrm{loc}}(B_1)$, $p > n$, be a strong solution of $\Delta u = -\chi_{\{u>0\}}$ in B_1 with*

$$u(x_1, x_2) = M(x_1^2 - x_2^2) - \kappa, \qquad (x_1, x_2) \in \partial B_1,$$

obtained by Proposition 2.20. Then $u \notin C^{1,1}_{\mathrm{loc}}(B_1)$ if M is sufficiently large.

The proof requires the familiarity with the blowups and Weiss-type monotonicity formulas and is outlined in Exercise 2.8. The reader may come back to this proof after reading Chapter 3.

We would like to conclude with a remark that the function u in Proposition 2.21 has the property

$$\frac{u(rx)}{r^{(n-1)/2}\|u\|_{L^2(\partial B_r)}} \to \frac{x_1^2 - x_2^2}{\|x_1^2 - x_2^2\|_{L^2(\partial B_1)}} \qquad \text{as} \quad r \to 0,$$

in $C^{1,\alpha}_{\mathrm{loc}}(\mathbb{R}^n)$, which means that the origin is a "cross-shaped singularity;" see Fig. 2.5. For details, we refer to the original paper [**AW06**]. It is also worth mentioning that the application of Proposition 2.20 gives a simple existence of so-called degenerate free boundary points in the unstable obstacle problem, where the solution decays faster than quadratically at a free boundary point. This, however, cannot happen in our model problems **A**, **B**, and **C**, which is proved in the beginning of the next chapter.

Notes

The optimal regularity in the classical obstacle problem is due to Frehse [**Fre72**]. For second order linear operators, similar results are due to Brezis-Kinderlehrer [**BK74**] and Gerhardt [**Ger73**]. Monneau [**Mon09**] considers the case when the right-hand side f in the classical obstacle problem is continuous only at a point with a Dini modulus of continuity in an averaged (L^p) sense.

The optimal regularity in the no-sign obstacle problem in dimension 2 (with $f = 1$) is due to Sakai [**Sak91**]. In higher dimensions the result is due to Caffarelli-Karp-Shahgholian [**CKS00**] (with $f = 1$), Shahgholian [**Sha03**] (with f Lipschitz). Petrosyan-Shahgholian [**PS07**], using a combination of two monotonicity formulas, proved the optimal regularity of the solution subject to a double-Dini condition on f. The most complete work in this regard is due to Andersson-Lindgren-Shahgholian [**ALS12**]. Using harmonic analysis methods, and no monotonicity formula, the authors give a complete proof of $C^{1,1}$ regularity of the no-sign obstacle problem with a Dini right-hand side f.

Optimal regularity for the two-phase problem was first proved by Uraltseva in [**Ura01**] by using the ACF monotonicity formula. Earlier, the quadratic growth near branch points (see Definition 3.24) was proved by Weiss [**Wei01**], by using a combination of monotonicity formulas of Weiss (see §3.5) and Almgren (see §9.3).

The proof of the $C^{1,1}$ regularity of solutions of obstacle-type problems in §2.3 is due to Shahgholian [**Sha03**]. The optimal regularity in the parabolic counterpart of Problem **A** is proved in Caffarelli-Petrosyan-Shahgholian [**CPS04**] and for more general case in Edquist-Petrosyan [**EP08**]. For the parabolic version of Problem **C**, it is done in Shahgholian-Uraltseva-Weiss [**SUW09**].

The $C^{1,1}$ regularity up to the fixed boundary is due to Jensen [**Jen80**] for the classical obstacle problem. The method for treating general obstacle-type problems as in §2.4 is due to Uraltseva [**Ura96, Ura07**] and Apushkinskaya-Uraltseva [**AU07**]. For the parabolic counterparts, we refer to Apushkinskaya-Shahgholian-Uraltseva [**ASU00**] and Apushkinskaya-Uraltseva [**AU09**].

The counterexample in §2.5 (due to Andersson-Weiss [**AW06**]) gives an interesting new aspect to problems of this kind. The failure of optimal regularity in this case does not exclude the possibility of analyzing the free boundary. This is studied in the recent works by Andersson-Shahgholian-Weiss [**ASW10, ASW12**].

The ACF monotonicity formula first appeared in [**ACF84**] in the study of a two-phase free boundary problem, and since then has become one of the most important technical tools in the analysis of free boundaries. For a complete proof of the formula we refer to the book of Caffarelli-Salsa [**CS05**]. The case of equality as in Theorem 2.9 was already discussed in the original paper [**ACF84**, Lemma 6.6, Remark 6.1]; however, the proof in dimensions $n > 2$ was incomplete as it was not known at that time that the inequality $\alpha(\Sigma) \geq \alpha(\Sigma^*)$ is strict unless Σ is a spherical cap: the result of Kawohl [**Kaw86**] was not yet available. We thank Mark Allen for bringing this up to our attention.

There is also an equally important parabolic version of the ACF monotonicity formula, discovered by Caffarelli [**Caf93**]. Besides, various extensions were proved to allow better flexibility in applications. They are known in the literature as almost monotonicity formulas and typically come in the form of estimates. Here we mention some of those extensions. Caffarelli-Kenig [**CK98**] proved a parabolic formula for the operators in divergence form with Dini coefficients. Caffarelli-Jerison-Kenig [**CJK02**] extended the elliptic formula to allow nonzero right-hand sides (see Theorem 2.11). The parabolic version of [**CJK02**] has been established by Edquist-Petrosyan [**EP08**]. The most complete extension is due to Matevosyan-Petrosyan [**MP11**], where the authors proved almost monotonicity formulas for elliptic and parabolic equations in divergence form with double-Dini coefficients and nonzero right-hand sides.

Exercises

2.1. This exercise outlines an alternative proof of the quadratic growth for the solutions of the classical obstacle problem (Theorem 2.1).

(i) Let $u \in C^1(B_R)$ be a weak solution of $\Delta u = g$ in B_R with $g \leq M$ and assume additionally that $u(0) = 0$. Prove that

$$\int_{B_\rho} u \leq C_n M \rho^{n+2}, \quad 0 < \rho \leq R,$$

where $C_n = |B_1|/2(n+2)$. (This estimate is due to H. S. Shapiro.) *Hint:* The function $u(x) - M|x|^2/2n$ is superharmonic, thus its average on any sphere is nonpositive. Integration with respect to the radius gives the result.

(ii) Use (i) to give an alternative proof of Theorem 2.1 when $f \geq 0$. *Hint:* Note that u is subharmonic and use the L^∞-L^1 estimate

$$\sup_{B_r(x^0)} u \leq \frac{C_n}{r^n} \int_{B_{2r}(x^0)} u,$$

which is a direct corollary of the sub-mean-value property.

2.2. Show that if $\Delta v = \Delta \psi$ in B_{2R} and $\psi \in C^{1,1}(B_{2R})$, then

$$\|D^2 v\|_{L^\infty(B_R)} \leq C_n \left(\frac{\|v\|_{L^\infty(B_{2R})}}{R^2} + \|D^2\psi\|_{L^\infty(B_{2R})} \right).$$

Hint: One can assume that $\psi(x^0) = |\nabla\psi(x^0)| = 0$ by subtracting a linear function if necessary. Then the difference $w = v - \psi$ is a harmonic function, and the estimate above is obtained from the corresponding estimate for w.

2.3. Let $v \in C(D)$ be a nonnegative subharmonic function in an open set D of $\mathbb{R}^n$. Prove that $v \in W^{1,2}_{\mathrm{loc}}(D)$.

Hint: Mollifications v_ε of v satisfy the inequality

$$\int_D \nabla v_\varepsilon \cdot \nabla\phi \, dx \leq 0, \quad \text{for any } \phi \in C_0^\infty(D),\ \phi \geq 0.$$

Take $\phi = v_\varepsilon \zeta^2$ with $\zeta \in C_0^\infty(D)$ and let $\varepsilon \to 0+$.

2.4. Let v be as in Exercise 2.3, $D = B_1$. Prove the inequality

$$I(\rho) := \int_{B_\rho} \frac{|\nabla v(x)|^2}{|x|^{n-2}} \, dx \leq \frac{C_n}{(1-\rho)^2} \int_{B_1 \setminus B_\rho} v^2 \, dx,$$

for any $\rho \in [\frac{1}{2}, 1)$.

Hint: In the inequality

$$\int_{B_1} \nabla v \cdot \nabla \phi \, dx \le 0,$$

which holds for any nonnegative $\phi \in W_0^{1,2}(B_1)$ by Exercise 2.3, plug $\phi = v\hat{G}\zeta^2$, where $G(x) = |x|^{2-n}$, $\hat{G} = \min\{G, \delta^{2-n}\}$ for a small positive $\delta < \rho$ and $\zeta \in C_0^\infty(B_1)$ with $\zeta = 1$ on B_ρ and $|\nabla\zeta| \le C/(1-\rho)$ on B_1. Then argue as at the end of §2.2.1.

2.5. Prove that under the conditions of Exercise 2.4, $I(r)$ is an absolutely continuous function on $(0,1)$ and

$$I'(r) = \frac{1}{r^{n-2}} \int_{\partial B_r} |\nabla v|^2 \, dH^{n-1} \quad \text{for a.e. } r \in (0,1).$$

Hint: Use Exercises 2.3 and 2.4.

2.6. Let $w \in W^{1,2}_{\mathrm{loc}}(D) \cap C(D)$ be nonnegative in an open set D in $\mathbb{R}^n$. Prove that if $\Delta w \ge -a$ in the sense of distribution on $\{w > 0\}$ for some $a \ge 0$, then $\Delta w \ge -a$ in D.

Hint: Consider the inequality

$$\int_D \nabla w \cdot \nabla \eta \, dx \le a \int_D \eta,$$

where $\eta \ge 0$, $\eta \in W_0^{1,2}(E)$, $E := \{w > 0\}$, and plug $\eta = \psi_\varepsilon(w)\phi$ with $\phi \in C_0^\infty(D)$ and ψ_ε satisfying

$$0 \le \psi_\varepsilon \le 1, \quad \psi_\varepsilon' \ge 0, \quad \psi_\varepsilon(t) = 0 \text{ for } t \le \varepsilon/2, \quad \psi_\varepsilon(t) = 1 \text{ for } t \ge \varepsilon.$$

2.7. Prove the Friedland-Hayman inequality (see the end of §2.4) in dimension $n = 2$, thereby completing the proof of Theorem 2.4 in dimension $n = 2$.

2.8. Let u be the solution of the unstable obstacle problem from Proposition 2.21. This exercise gives the proof that $u \notin C^{1,1}(B_{1/2})$. It assumes familiarity with the methods from §3.5.

1) Show that the Weiss-type functional

$$W(r,u) = \frac{1}{r^{n+2}} \int_{B_r} (|\nabla u|^2 - 2u^+) dx - \frac{2}{r^{n+3}} \int_{\partial B_r} u^2 dH^{n-1}$$

is monotone nondecreasing for $r \in (0,1)$ and

$$\frac{d}{dr} W(r,u) = \frac{2}{r^{n+4}} \int_{\partial B_r} (x \cdot \nabla u - 2u)^2.$$

2) Show that if $u \in C^{1,1}(B_{1/2})$, then any blowup u_0 of u at the origin, i.e., the limit of rescalings

$$u_r(x) = \frac{u(rx)}{r^2}$$

over any sequence $r = r_j \to 0$, is a homogeneous of degree 2 solution of the unstable obstacle problem.

3) In dimension $n = 2$, show that the only homogeneous of degree 2 solution of the unstable obstacle problem, which is also from the class $C^0_{*2}(B_1)$, is the identically zero function. In fact, without the symmetry assumption, up to a rotation, there exists only one nonzero homogeneous solution; this solution will be in $C^0_{*3}(B_1)$.

4) By a direct computation show that $W(1, u) < 0$ if the constant M in Proposition 2.21 is sufficiently large. Hence, if $u \in C^{1,1}(B_{1/2})$, then for any blowup u_0 at the origin,

$$W(1, u_0) = W(0+, u) < 0,$$

which means that u_0 is nonzero. However, this contradicts the result in part 3) above, which completes the proof of Proposition 2.21.

Chapter 3

Preliminary analysis of the free boundary

In this chapter we establish some basic properties of solutions of our model Problems **A**, **B**, **C**. In §3.1 we prove the nondegeneracy property of solutions, which says that, in a sense, the solutions u do not decay faster than quadratically near free boundary points. This property is then used to show that the free boundary $\Gamma(u)$ has a Lebesgue measure zero. A more accurate analysis shows that actually $H^{n-1}(\Gamma(u)) < \infty$ (§3.2). Next, in §3.3 we introduce the method of rescalings and blowups. We then prove a general theorem for blowups in obstacle-type problems in §3.4 and classify them for Problems **A**, **B**, **C**. This in turn gives a classification of free boundary points. Finally, in §3.5, we introduce Weiss's monotonicity formula, which is another useful tool in the study of blowups, especially in Problems **A** and **C**.

3.1. Nondegeneracy

We start with an important nondegeneracy property.

By the results of the previous chapter, for solutions of Problem **A**, we have the quadratic growth estimate

$$\sup_{B_r(x^0)} u \leq u(x^0) + \frac{M}{2}r^2,$$

for any $x^0 \in D$, as long as $B_r(x^0) \subset D$, where $M = \|D^2u\|_{L^\infty(D)}$. However, this estimate alone does not exclude the possibility that $u(x) - u(x^0)$ may decay faster than quadratically at x^0. We next show that this does not happen when $x^0 \in \overline{\Omega(u)}$ and in particular when $x^0 \in \Gamma(u)$. Similar nondegeneracy statements hold also for Problems **B** and **C**.

Lemma 3.1 (Nondegeneracy: Problem **A**). *Let u be a solution of Problem* **A** *in D. Then we have the inequality*

$$\sup_{\partial B_r(x^0)} u \geq u(x^0) + \frac{r^2}{8n}, \quad \textit{for any } x^0 \in \overline{\Omega(u)}, \tag{3.1}$$

provided $B_r(x^0) \Subset D$.

Remark 3.2. Since the solutions of Problem **A** are subharmonic, we can replace sup over $\partial B_r(x^0)$ to the one over $B_r(x^0)$, obtaining an equivalent statement.

Before giving the proof of Lemma 3.1, we would like to illustrate one of its main ideas by proving a similar nondegeneracy statement for solutions of $\Delta u = 1$.

Lemma 3.3. *Let u satisfy $\Delta u = 1$ in the ball B_R. Then*

$$\sup_{\partial B_r} u \geq u(0) + \frac{r^2}{2n}, \quad 0 < r < R.$$

Proof. Consider the auxiliary function

$$w(x) = u(x) - \frac{|x|^2}{2n}, \quad x \in B_R.$$

Then w is harmonic in B_R. Therefore by the maximum principle we obtain that

$$w(0) \leq \sup_{\partial B_r} w = \left(\sup_{\partial B_r} u\right) - \frac{r^2}{2n},$$

which implies the required inequality. □

Proof of Lemma 3.1. 1) Assume first that $x^0 \in \Omega(u)$ and moreover $u(x^0) > 0$. Consider then the auxiliary function

$$w(x) = u(x) - u(x^0) - \frac{|x - x^0|^2}{2n}, \tag{3.2}$$

similar to the one in the proof of the previous lemma. We have $\Delta w = 0$ in $B_r(x^0) \cap \Omega$. Since $w(x^0) = 0$, by the maximum principle we have that

$$\sup_{\partial(B_r(x^0) \cap \Omega)} w \geq 0.$$

Besides, $w(x) = -u(x^0) - |x - x^0|^2/(2n) < 0$ on $\partial\Omega$. Therefore, we must have

$$\sup_{\partial B_r(x^0) \cap \Omega} w \geq 0.$$

The latter is equivalent to

$$\sup_{\partial B_r(x^0)\cap\Omega} u \geq u(x^0) + \frac{r^2}{2n},$$

and the lemma is proved in this case.

2) Suppose now that $x^0 \in \Omega(u)$ and $u(x^0) \leq 0$. If $B_{r/2}(x^0)$ contains a point x^1 such that $u(x^1) > 0$, then

$$\sup_{B_r(x^0)} u \geq \sup_{B_{r/2}(x^1)} u \geq u(x^1) + \frac{(r/2)^2}{2n} \geq u(x^0) + \frac{r^2}{8n}$$

and we finish the proof by recalling Remark 3.2. If it happens that $u \leq 0$ in $B_{r/2}(x^0)$, from subharmonicity of u and the strong maximum principle we will have that either $u = 0$ identically in $B_{r/2}(x^0)$ or $u < 0$ in $B_{r/2}(x^0)$. The former case is impossible, as $x^0 \in \Omega(u)$, and the latter case implies that $B_{r/2}(x^0) \subset \Omega(u)$ and therefore $\Delta u = 1$ in $B_{r/2}(x^0)$. Then Lemma 3.3 finishes the proof in this case, and we obtain

$$\sup_{B_r(x^0)} u \geq \sup_{B_{r/2}(x^0)} u \geq u(x^0) + \frac{r^2}{8n}.$$

3) Finally, for $x^0 \in \overline{\Omega(u)}$, we take a sequence $x^j \in \Omega(u)$ such that $x^j \to x^0$ as $j \to \infty$ and pass to the limit in the corresponding nondegeneracy statement at x^j. □

Even though the proof above does not work for Problem **B** in general, we still have the nondegeneracy. This also gives an alternative proof of Lemma 3.1.

Lemma 3.4 (Nondegeneracy: Problem **B**). *Let u be a solution of Problem* **B** *in D. Then we have the inequality*

$$\sup_{\partial B_r(x^0)} u \geq u(x^0) + \frac{r^2}{2n}, \quad \text{for any } x^0 \in \overline{\Omega(u)}, \tag{3.3}$$

provided $B_r(x^0) \Subset D$.

Proof. By continuity, it suffices to obtain the estimate (3.3) only for points $x^0 \in \Omega(u)$. Note that at those points we have $|\nabla u(x^0)| \neq 0$. Consider now the same auxiliary function w as in (3.2). Then we claim that

$$\sup_{B_r(x^0)} w = \sup_{\partial B_r(x^0)} w. \tag{3.4}$$

Indeed, suppose the supremum of w is attained at some interior point $y \in B_r(x^0)$, implying that $|\nabla w(y)| = 0$. The latter is equivalent to

$$|\nabla u(y)| = \frac{|y - x^0|}{n}.$$

In particular, $y \neq x^0$, since otherwise $|\nabla u(x^0)| = 0$, contrary to our assumption. Hence $|\nabla u(y)| > 0$ and consequently $y \in \Omega(u)$. Since w is harmonic in $\Omega(u)$, by the strong maximum principle w is constant in a neighborhood of y. Thus, the set of maxima of w is both relatively open and closed in $B_r(x^0)$, which implies that w is constant there and (3.4) is trivially satisfied.

To finish the proof observe that (3.4) easily implies (3.3):

$$0 = w(x^0) \leq \sup_{\partial B_r(x^0)} w = \sup_{\partial B_r(x^0)} u - \frac{r^2}{2n} - u(x^0). \qquad \square$$

Remark 3.5. We would like to emphasize that for the proof of nondegeneracy in Lemmas 3.1 and 3.4 we have only used that $\Delta u \geq 1$ in $\Omega(u)$ and $\Delta u \geq 0$ in D. The equation itself is not needed. The same remark holds also for Lemma 3.10, Corollary 3.11, and Lemma 3.12.

Finally, in Problem **C** we have nondegeneracy in both phases, provided $\lambda_\pm > 0$.

Lemma 3.6 (Nondegeneracy: Problem **C**). *If u is a solution of Problem* **C** *in D, then we have*

$$\sup_{\partial B_r(x^0)} u \geq u(x^0) + \lambda_+ \frac{r^2}{2n}, \quad \text{for any } x^0 \in \overline{\Omega_+(u)}, \tag{3.5}$$

$$\inf_{\partial B_r(x^0)} u \leq u(x^0) - \lambda_- \frac{r^2}{2n}, \quad \text{for any } x^0 \in \overline{\Omega_-(u)}, \tag{3.6}$$

provided $B_r(x^0) \Subset D$.

Proof. To prove these inequalities, we consider the auxiliary functions

$$w(x) = u(x) - u(x^0) \mp \lambda_\pm \frac{|x - x^0|^2}{2n}$$

and argue similarly to part 1) of the proof of Lemma 3.1. We leave the details to the reader. $\square$

Corollary 3.7 (Nondegeneracy of the gradient). *Under the conditions of either Lemmas* 3.1, 3.4, *or* 3.6 *the following inequality holds:*

$$\sup_{B_r(x^0)} |\nabla u| \geq c_0 r,$$

for a positive c_0, depending only on n in Problems **A**, **B**, *and also on $\lambda_\pm$ for Problem* **C**.

The proof is left as an exercise to the reader.

3.2. Lebesgue and Hausdorff measures of the free boundary

3.2.1. Porosity of the free boundary.

Definition 3.8. We say that a measurable set $E \subset \mathbb{R}^n$ is *porous* with a porosity constant $0 < \delta < 1$ if every ball $B = B_r(x)$ contains a smaller ball $B' = B_{\delta r}(y)$ such that

$$B_{\delta r}(y) \subset B_r(x) \setminus E.$$

We say that E is *locally porous* in an open set D if $E \cap K$ is porous (with possibly different porosity constants) for any $K \Subset D$.

It is clear that the Lebesgue upper density of a porous set E at $x^0 \in E$ is less than one:

$$\limsup_{r \to 0} \frac{|E \cap B_r(x^0)|}{|B_r|} \leq 1 - \delta^n < 1.$$

This has an immediate corollary.

Proposition 3.9. *If $E \subset \mathbb{R}^n$ is porous, then $|E| = 0$. If E is locally porous in D, then $|E \cap D| = 0$.*

Proof. The proof follows from the fact that the Lebesgue upper density must be 1 for almost all points of E and the remark right before the proposition. □

Returning to the study of our model problems, we next observe that the combination of the nondegeneracy and the $C^{1,1}$ regularity naturally implies the porosity of the free boundary.

Lemma 3.10 (Porosity of the free boundary). *Let u be a solution of Problem* **A** *or* **B** *in an open set $D \subset \mathbb{R}^n$. Then $\Gamma(u)$ is locally porous in D.*

If u is a solution of Problem **C**, *then $\Gamma^0(u) = \Gamma(u) \cap \{|\nabla u| = 0\}$ is locally porous.*

Proof. For Problems **A**, **B**, let $K \Subset D$, $x^0 \in \Gamma(u)$ and $B_r(x^0) \subset K$. Using the nondegeneracy of the gradient (see Corollary 3.7), one can find $y \in \overline{B_{r/2}(x^0)}$ such that

$$|\nabla u(y)| \geq \frac{c_0}{2} r.$$

Now, using that $M = \|D^2 u\|_{L^\infty(K)} < \infty$, we will have

$$\inf_{B_{\delta r}(y)} |\nabla u| \geq \left(\frac{c_0}{2} - M\delta\right) r \geq \frac{c_0}{4} r \quad \text{if } \delta = \frac{c_0}{4M}.$$

This implies that

$$B_{\tilde{\delta} r}(y) \subset B_r(x^0) \cap \Omega(u) \subset B_r(x^0) \setminus \Gamma,$$

where $\tilde{\delta} = \min\{\delta, 1/2\}$. Hence the porosity condition is satisfied for any ball centered at $\Gamma(u)$ and contained in K. This is sufficient to conclude that $\Gamma(u) \cap K$ is porous; see Exercise 3.4. Thus, $\Gamma(u)$ is locally porous in D.

For Problem **C**, the same argument as above shows that

$$B_{\tilde{\delta}r}(y) \subset B_r(x^0) \cap [\Omega(u) \cup \Gamma^*(u)] \subset B_r(x^0) \setminus \Gamma^0(u),$$

which implies the local porosity of $\Gamma^0(u)$. □

Corollary 3.11 (Lebesgue measure of Γ). *Let u be a solution of Problem* **A**, **B**, *or* **C** *in D. Then $\Gamma(u)$ has a Lebesgue measure zero.*

Proof. In the case of Problems **A**, **B** the statement follows immediately from the local porosity of $\Gamma(u)$ and Proposition 3.9.

In the case of Problem **C**, we obtain $|\Gamma^0(u)| = 0$. On the other hand, $\Gamma^*(u)$ is locally a $C^{1,\alpha}$ surface and therefore also has a Lebesgue measure zero. Hence, $|\Gamma(u)| = 0$ in this case as well. □

We finish this subsection with the following observation.

Lemma 3.12 (Density of Ω). *Let u be a solution of Problem* **A**, **B**, *or* **C** *in D and $x^0 \in \Gamma(u)$. Then*

$$\frac{|B_r(x^0) \cap \Omega(u)|}{|B_r|} \geq \beta, \tag{3.7}$$

provided $B_r(x^0) \subset D$, where β depends only on $\|D^2 u\|_{L^\infty(D)}$ and n for Problems **A**, **B** *and additionally on $\lambda_\pm$ for Problem* **C**.

Proof. The proof of Lemma 3.10 shows that

$$\frac{|B_r(x^0) \cap \Omega(u)|}{|B_r|} \geq \tilde{\delta}^n$$

in the case of Problems **A** and **B**, and that

$$\frac{|B_r(x^0) \cap \Omega(u)|}{|B_r|} = \frac{|B_r(x^0) \cap [\Omega(u) \cup \Gamma^*(u)]|}{|B_r|} \geq \tilde{\delta}^n$$

in the case of Problem **C**. This completes the proof. □

3.2.2. $(n-1)$-Hausdorff measure of the free boundary. The porosity of the free boundary implies not only that its Lebesgue measure is zero but also that it actually has a Hausdorff dimension less than n (see e.g. [**Sar75**]). Here, however, we will prove a stronger result.

Lemma 3.13 ($(n-1)$-Hausdorff measure of Γ). *If u is a $C^{1,1}$ solution of Problem* **A**, **B**, *or* **C** *in an open set $D \subset \mathbb{R}^n$, then $\Gamma(u)$ is a set of finite $(n-1)$-dimensional Hausdorff measure locally in D.*

Proof. Let

$$v_i = \partial_{x_i} u, \quad i = 1, \dots, n, \qquad E_\varepsilon = \{|\nabla u| < \varepsilon\} \cap \Omega(u).$$

Observe that

$$c_0 \le |\Delta u|^2 \le c_n \sum_{i=1}^n |\nabla v_i|^2 \quad \text{in } \Omega,$$

where $c_0 = 1$ in the case of Problems **A**, **B** and $c_0 = \min\{\lambda_+^2, \lambda_-^2\}$ for Problem **C**. Thus, for an arbitrary compact set $K \Subset D$ we have

$$c_0\,|K \cap E_\varepsilon| \le c_n \int_{K\cap E_\varepsilon} \sum_i |\nabla v_i|^2 dx \le c_n \sum_i \int_{K\cap\{|v_i|<\varepsilon\}\cap\Omega(u)} |\nabla v_i|^2 dx.$$

To estimate the right-hand side, we apply Lemma 2.15, noticing that we can take $M_1 = M_2 = 0$ for solutions of Problems **A**, **B**, and **C**. This gives

$$\int_D \nabla v_i^\pm \nabla \eta \, dx \le 0, \quad i = 1, \dots, n$$

for any nonnegative $\eta \in C_0^\infty(D)$, and by continuity, for any nonnegative $\eta \in W_0^{1,2}(D)$. If we now choose $\eta = \psi_\varepsilon(v_i^\pm)\phi$, with

$$\psi_\varepsilon(t) = \begin{cases} 0, & t \le 0, \\ \varepsilon^{-1} t, & 0 \le t \le \varepsilon, \\ 1, & t \ge \varepsilon, \end{cases}$$

and a nonnegative cutoff function $\phi \in C_0^\infty(D)$, $\phi = 1$ on K, we obtain

$$\begin{aligned} &\int_D \nabla v_i^\pm \nabla(\psi_\varepsilon(v_i^\pm)\phi) dx \\ &\quad = \int_{\{0<v_i^\pm<\varepsilon\}} \varepsilon^{-1} |\nabla v_i^\pm|^2 \phi \, dx + \int_D \nabla v_i^\pm \psi_\varepsilon(v_i^\pm) \nabla\phi \, dx \le 0, \end{aligned}$$

which implies that

$$\begin{aligned} \varepsilon^{-1} \int_{K\cap\{|v_i|<\varepsilon\}\cap\Omega(u)} |\nabla v_i|^2 dx &\le \varepsilon^{-1} \int_{\{0<|v_i|<\varepsilon\}} |\nabla v_i|^2 \phi \, dx \\ &\le \int_D |\nabla v_i| |\nabla\phi| \, dx \le c_n M \int_D |\nabla\phi| \, dx, \end{aligned}$$

where $M = \|D^2 u\|_{L^\infty(D)}$. Thus, summing over $i = 1, \dots, n$, we arrive at an estimate

$$c_0\,|K \cap E_\varepsilon| \le C\varepsilon M, \tag{3.8}$$

where $C = C(n, K, D)$.

Consider now a covering of $\Gamma \cap K$ by a finite family $\{B^i\}_{i\in I}$ of balls of radius ε centered on $\Gamma \cap K$, such that no more than $N = N_n$ balls from this family overlap (such covering exists because of the Besicovitch covering lemma). For $\varepsilon > 0$ sufficiently small, we may assume that $B^i \subset K'$ for a slightly larger compact K' so that $K \Subset \mathrm{Int}(K') \Subset D$. For Problems **A** and **B**, now notice that $|\nabla u| < M\varepsilon$ in each B^i, implying that $B^i \cap \Omega \subset E_{M\varepsilon}$. Then, using (3.7) and (3.8), we obtain

$$\begin{aligned}\sum_{i\in I}|B^i| &\le \frac{1}{\beta}\sum_{i\in I}|B^i\cap\Omega| \le \frac{1}{\beta}\sum_{i\in I}|B^i\cap E_{M\varepsilon}|\\ &\le \frac{N}{\beta}|K'\cap E_{M\varepsilon}| \le \frac{CNM^2\varepsilon}{c_0\beta}.\end{aligned}$$

This gives the estimate

$$\sum_{i\in I}\mathrm{diam}(B^i)^{n-1} \le C(n,M,K,D),$$

and, letting $\varepsilon \to 0$, we conclude that

$$H^{n-1}(\Gamma(u)\cap K) \le C(n,M,K,D).$$

This completes the proof of the lemma for Problems **A** and **B**.

For Problem **C**, the same proof as above works for $\Gamma^0(u)$. To estimate $H^{n-1}(\Gamma^*(u))$, we need to fully use the equation for the directional derivative $v = \partial_e u$, and not only the fact that $v_\pm$ are subharmonic. First, note that for any $\eta \in C_0^\infty(D)$ we have

$$\int_D \nabla v\cdot\nabla\eta = \int_D \left(\lambda_+\chi_{\{u>0\}} - \lambda_-\chi_{\{u<0\}}\right)\partial_e\eta.$$

In the case when $\eta \in C_0^\infty(D\setminus\Gamma^0(u))$, the right-hand side can be written as

$$-(\lambda_+ + \lambda_-)\int_{\Gamma^*(u)}(e\cdot\nu)\eta\, dH^{n-1},$$

where $\nu = \nabla u(x)/|\nabla u(x)|$. By continuity, this can be extended to $\eta \in W_0^{1,2}(D\setminus\Gamma^0(u))$. Now take $\eta = \psi_\varepsilon(v)\phi$ with the same ϕ, ψ_ε as before. It is easy to see that $\eta \in W_0^{1,2}(D)$ and that $\eta = 0$ a.e. on $\Gamma^0(u)$, but one still has to show that η is indeed in $W^{1,2}(D\setminus\Gamma^0(u))$. This is indicated in Exercise 3.7. Then using the same computations as in the derivation of (3.8) we obtain

$$\begin{aligned}\varepsilon^{-1}\int_{K\cap\{0<v<\varepsilon\}}|\nabla v|^2dx + (\lambda_+ + \lambda_-)\int_{\Gamma^*(u)\cap K}(e\cdot\nu)\psi_\varepsilon(v)dH^{n-1}\\ \le c_nM\int_D|\nabla\phi|dx.\end{aligned}$$

Next, observe that $v = \partial_e u > 0$ for $\psi_\varepsilon \ne 0$; therefore $e\cdot\nu > 0$ in the integral over $\Gamma^*(u)\cap K$. This implies that both integrals on the left-hand side are nonnegative and therefore

$$(\lambda_+ + \lambda_-)\int_{\Gamma^*(u)\cap K}(e\cdot\nu)\psi_\varepsilon(v)dH^{n-1} \le CM.$$

Letting $\varepsilon \to 0$, we arrive at the estimate

$$(\lambda_+ + \lambda_-) \int_{\Gamma^*(u)\cap K} (e \cdot \nu)_+ dH^{n-1} \leq CM,$$

for any direction e in $\mathbb{R}^n$. Choosing $e = e_1, \ldots, e_n$, and observing that $e_i \cdot \nu \geq 1/\sqrt{n}$ for at least one $i = 1, \ldots, n$, one obtains that

$$H^{n-1}(\Gamma^*(u) \cap K) \leq \frac{CM}{\lambda_+ + \lambda_-}. \qquad \square$$

Remark 3.14. The estimate (3.8) essentially means that

$$|\Omega \cap \{|\nabla u| < \varepsilon\}| \leq C\varepsilon.$$

In particular, it implies that $|\Omega \cap \{|\nabla u| = 0\}| = 0$.

3.3. Classes of solutions, rescalings, and blowups

3.3.1. Local and global solutions. The further analysis of the free boundary is based on the study of so-called blowups. Since the regularity of the free boundary is a local question, we may restrict ourselves to solutions defined in balls centered at free boundary points. We start with the definition of the appropriate classes of normalized solutions.

Definition 3.15 (Local solutions). For given $R, M > 0$, and $x^0 \in \mathbb{R}^n$ let $P_R(x^0, M)$ be the class of $C^{1,1}$ solutions u of Problems **A**, **B**, or **C** in $B_R(x^0)$ such that

- $\|D^2 u\|_{L^\infty(B_R(x^0))} \leq M$;
- $x^0 \in \Gamma(u)$ in Problems **A**, **B**; $x^0 \in \Gamma^0(u)$ in Problem **C**.

We will use the abbreviated notation $P_R(M)$ for the class $P_R(0, M)$.

Taking formally $R = \infty$ in the above definition, we obtain solutions in the entire space $\mathbb{R}^n$, which grow quadratically at infinity. Slightly abusing the terminology, we call them *global solutions.*

Definition 3.16 (Global solutions). For given $M > 0$ and $x^0 \in \mathbb{R}^n$ let $P_\infty(x^0, M)$ be the class of $C^{1,1}_{\mathrm{loc}}$ solutions u of Problems **A**, **B**, or **C** in $\mathbb{R}^n$, such that

- $\|D^2 u\|_{L^\infty(\mathbb{R}^n)} \leq M$;
- $x^0 \in \Gamma(u)$ in Problems **A**, **B**; $x^0 \in \Gamma^0(u)$ in Problem **C**.

The class $P_\infty(0, M)$ will also be denoted by $P_\infty(M)$.

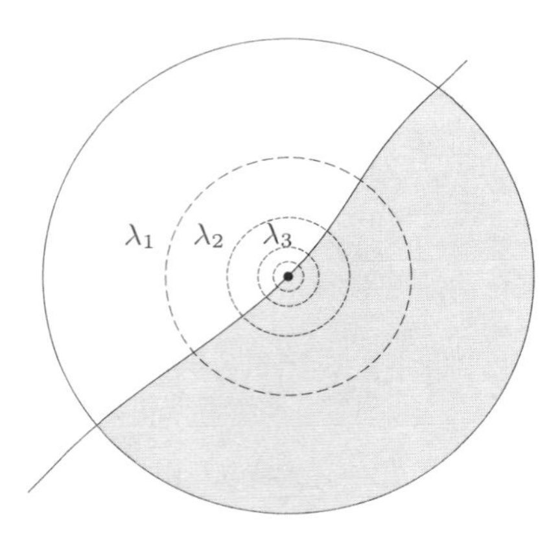

Figure 3.1. Blowup over the sequence of radii $\lambda_j \to 0$

3.3.2. Rescalings and blowups. The following scaling and translation properties are enjoyed by the solutions in the above classes. If $u \in P_R(x^0, M)$ and $\lambda > 0$, then the *rescaling* of u at x^0,

$$u_\lambda(x) = u_{x^0,\lambda}(x) := \frac{u(x^0 + \lambda x) - u(x^0)}{\lambda^2}, \quad x \in B_{R/\lambda}, \tag{3.9}$$

will be from the class $P_{R/\lambda}(M)$. Using this simple observation, we will often state the results for normalized classes $P_R(M)$ or even $P_1(M)$ as the corresponding statements for classes $P_R(x^0, M)$ can be easily recovered.

Observe that for $u \in P_R(M)$ the rescalings u_λ satisfy the estimate $|D^2 u_\lambda(x)| \leq M$ in $B_{R/\lambda}$ for all $\lambda > 0$. Therefore, we also have

$$|\nabla u_\lambda(x)| \leq M|x|, \quad x \in B_{R/\lambda},$$
$$|u_\lambda(x)| \leq \frac{1}{2} M|x|^2, \quad x \in B_{R/\lambda}.$$

Hence, we can find a sequence $\lambda = \lambda_j \to 0$ such that

$$u_{\lambda_j} \to u_0 \quad \text{in } C^{1,\alpha}_{\mathrm{loc}}(\mathbb{R}^n) \text{ for any } 0 < \alpha < 1,$$

where $u_0 \in C^{1,1}_{\mathrm{loc}}(\mathbb{R}^n)$. (See Figs. 0.2 and 3.1 for illustration.) Such u_0 is called a *blowup of* u *at* x^0, and Proposition 3.17 below implies that u_0 is a global solution; more precisely, $u_0 \in P_\infty(M)$. An important remark is that it is a priori not clear if u_0 is unique, as different sequences $\lambda_j \to 0$ may lead to different limits u_0.

The construction above can be easily generalized to the case when we have a sequence of free boundary points $x^j \to x^0$ (variable centers) instead of a fixed center x^0. Namely, we consider the limits

$$u_{x^j,\lambda_j} \to u_0 \quad \text{in } C^{1,\alpha}_{\mathrm{loc}}(\mathbb{R}^n) \text{ for any } 0 < \alpha < 1$$

as $\lambda_j \to 0$. We call such u_0 *blowups over the sequence* $x^j \to x^0$ (or blowups with variable centers). Furthermore, sometimes we may need to study the limits of rescalings $(u_j)_{x^j,\lambda_j}$ for different $u_j \in P_R(x^j, M)$, and we will still call such limits blowups.

In the case when $u \in P_\infty(M)$, we may also let $\lambda \to \infty$. The resulting limits are called *shrinkdowns* (or *blowdowns*), instead of blowups.

Proposition 3.17 (Limits of solutions). *Let $\{u_j\}_{j=1}^\infty$ be a sequence of solutions to Problem* **A**, **B**, *or* **C** *in an open set D, such that*

$$u_j \to u_0 \quad \text{in } C^{1,\alpha}_{\text{loc}}(D),$$

for some $0 < \alpha < 1$. Then the following statements hold:

(i) *If $x^0 \in D$, then we have the implications*

$$\begin{aligned} u_0(x^0) > 0 \quad &\Rightarrow \quad u_j > 0 \text{ on } B_\delta(x^0), \quad j \geq j_0,\\ u_0(x^0) < 0 \quad &\Rightarrow \quad u_j < 0 \text{ on } B_\delta(x^0), \quad j \geq j_0,\\ |\nabla u_0(x^0)| > 0 \quad &\Rightarrow \quad |\nabla u_j| > 0 \text{ on } B_\delta(x^0), \quad j \geq j_0, \end{aligned}$$

for some $\delta > 0$ and sufficiently large j_0.

(ii) *If $B_\delta(x^0) \subset D$, then we have*

$$|\nabla u_0| = 0 \text{ on } B_\delta(x^0) \quad \Rightarrow \quad |\nabla u_j| = 0 \text{ on } B_{\delta/2}(x^0), \quad j \geq j_0,$$

for sufficiently large j_0.

(iii) *u_0 solves the same Problem* **A**, **B**, *or* **C**, *as u_j, $j = 1, 2, \dots$.*

(iv) *If $x^{j_k} \in \Gamma(u_{j_k})$, for some $j_k \to \infty$, and $x^{j_k} \to x^0 \in D$, then $x^0 \in \Gamma(u_0)$.*

(v) *$u_j \to u_0$ strongly in $W^{2,p}_{\text{loc}}(D)$ for any $1 < p < \infty$.*

Proof. (i) The implications in this part are immediate corollaries of the C^1_{loc} convergence $u_j \to u_0$.

(ii) Suppose this fails; then there exists $j_k \to \infty$ such that $|\nabla u_{j_k}(y^k)| > 0$ for some $y^k \in B_{\delta/2}(x^0)$ and $|\nabla u_0| = 0$ in $B_\delta(x^0)$. Applying the nondegeneracy of the gradient (Corollary 3.7) at y^k and using the fact that $B_{\delta/4}(y^k) \subset B_{(3/4)\delta}(x^0)$, we obtain

$$\sup_{B_{(3/4)\delta}(x^0)} |\nabla u_{j_k}| \geq c\delta.$$

Then, passing to the limit, we arrive at

$$\sup_{B_{(3/4)\delta}(x^0)} |\nabla u_0| \geq c\delta,$$

contradicting our assumption.

(iii) To show that u_0 solves the same Problem **A**, **B**, or **C**, note that without loss of generality, we may assume that $\{u_j\}$ is uniformly bounded in $W^{2,p}(K)$, $1 < p \leq \infty$, for any $K \Subset D$, and hence $u_0 \in W^{2,p}_{\mathrm{loc}}(D)$. Therefore it is enough to verify that the equation for u_0 is satisfied a.e. in D.

Consider first Problems **A**, **B**. Since $\nabla u_0 = 0$ on $\Omega^c(u_0)$, $\Delta u_0 = 0$ a.e. there. If $x^0 \in \Omega(u_0)$, then from (i) above we obtain that $B_\delta(x^0) \subset \Omega(u_j)$ for some $\delta > 0$ and $j \geq j_0$. Consequently

$$\Delta u_j = 1 \quad \text{in } B_\delta(x^0), \quad j \geq j_0,$$

which implies that

$$\Delta u_0 = 1 \quad \text{in } \Omega(u_0).$$

Thus,

$$\Delta u_0 = \chi_{\Omega(u_0)}, \quad \text{a.e. in } D,$$

as required.

In Problem **C**, we may use the same argument, since the portion $\Gamma^*(u_0)$ of the free boundary where $|\nabla u_0| \neq 0$ is a C^1 surface and thus has a Lebesgue measure zero. Hence, u_0 satisfies the desired equation.

(iv) If $x^{j_k} \in \Gamma(u_{j_k}) \subset \Omega^c(u_{j_k})$, the implications in (i) show that $x^0 \in \Omega^c(u_0)$. So the question is whether $x^0 \in \Gamma(u_0)$ or there exists a ball $B_\delta(x^0) \subset \Omega^c(u_0)$. Now note that in such a ball $|\nabla u_0| = 0$ in the case of all Problems **A**, **B**, or **C**. Thus, by part (ii), we will have that $B_{\delta/2}(x^{j_k}) \subset \Omega^c(u_{j_k})$, contradicting the fact that $x^{j_k} \in \Gamma(u_{j_k})$. Hence, we necessarily have $x^0 \in \Gamma(u_0)$.

(v) To prove that $u_j \to u_0$ strongly in $W^{2,p}_{\mathrm{loc}}$, $1 < p < \infty$, it will suffice to show that

$$D^2 u_j \to D^2 u_0 \quad \text{a.e. in } D,$$

since $D^2 u_j$ are uniformly bounded on any $K \Subset D$. Now, the pointwise convergence in $\Omega(u_0)$ follows from part (i). The pointwise convergence on $\mathrm{Int}(\Omega(u_0)^c)$ follows from part (ii). The only remaining points are the ones on the free boundary $\Gamma(u_0)$, and since it has a Lebesgue measure zero, we obtain the a.e. convergence of $D^2 u_j$ to $D^2 u_0$, as claimed. □

3.4. Blowups

The purpose of this section is to identify the blowups at free boundary points for solutions of Problems **A**, **B**, and **C**. The method that we are going to use works in the general case of functions satisfying the structural conditions **OT**$_1$–**OT**$_2$, with an additional condition that the rescalings converge strongly in $W^{2,p}$, for $p > n$.

Theorem 3.18 (Strong blowups: $\mathbf{OT}_1$–$\mathbf{OT}_2$). *Let $u \in C^{1,1}_{\rm loc}(D)$ satisfy $\mathbf{OT}_1$–$\mathbf{OT}_2$ and let $x^0 \in D$ be such that $|\nabla u(x^0)| = 0$. Then any strong limit*

$$u_0(x) = \lim_{j\to\infty} u_{x^0,r_j} \quad in \quad W^{2,p}_{\rm loc}(\mathbb{R}^n),\ p > n,$$

over a sequence $r_j \to 0+$ of rescalings

$$u_{x^0,r}(x) = \frac{u(x^0 + rx) - u(x^0)}{r^2}$$

is either a monotone function of one independent variable or a homogeneous quadratic polynomial.

Being a monotone function of one independent variable here means that there exist a unit vector e and a monotone function $\phi : \mathbb{R} \to \mathbb{R}$ such that $u_0(x) = \phi(x \cdot e)$ for all $x \in \mathbb{R}^n$.

Remark 3.19. Note that we explicitly require u_{x^0,r_j} to converge *strongly* to the function u_0 in the theorem above. Because of Proposition 3.17(v), this assumption is automatically satisfied in Problems **A**, **B**, and **C**.

To prove Theorem 3.18, we first identify the obstacle-type problem solved by blowups.

Lemma 3.20. *Let u and u_0 be as in Theorem 3.18. Also let $R > 0$ be such that $B_R(x^0) \subset D$. Then u_0 satisfies*

$$\Delta u_0 = a_+ \chi_{\{u_0>0\}} + a_- \chi_{\{u_0<0\}} \quad in \quad G_0,$$
$$|\nabla u_0| = 0 \quad in \quad \mathbb{R}^n \setminus G_0,$$

where

$$G_0 = \mathbb{R}^n \setminus \limsup_{j\to\infty}(B_{R/r_j} \setminus G(u_{r_j})),$$
$$a_+ = f(x^0, u(x^0)+), \quad a_- = f\left(x^0, u(x^0)-\right).$$

Here $G(u_r)$ is the appropriately translated and scaled version of $G(u)$ and $f(x^0, s\pm) = \lim_{\varepsilon\to0+} f(x^0, s \pm \varepsilon)$. Observe that $a_+ \geq a_-$ due to the second condition in $\mathbf{OT}_2$.

By $\limsup_{j\to\infty} E_j$ of the sequence of sets E_j in the lemma above we understand the set of all limit points $x = \lim_{j_k\to\infty} x^{j_k}$ for $x^{j_k} \in E_{j_k}$.

Proof. First, the rescaling u_r satisfies

$$\Delta u_r = f(x^0 + rx, u(x^0) + r^2 u_r) \quad \text{in} \quad G(u_r).$$

Next notice that the limits $f(x, s+)$ and $f(x, s-)$ exist for any $s \in \mathbb{R}$ due to the semimonotonicity of f in s, and that they are uniform in x. Now, if

$x \in G_0 \cap \{u_0 > 0\}$, then there exists a $\delta > 0$ such that $B_\delta(x) \subset G(u_{r_j}) \cap \{u_{r_j} > 0\}$ for sufficiently large j. Passing to the limit, we obtain that

$$\Delta u_0 = f(x^0, u(x^0)+) \quad \text{in} \quad G_0 \cap \{u_0 > 0\}.$$

Similarly,

$$\Delta u_0 = f(x^0, u(x^0)-) \quad \text{in} \quad G_0 \cap \{u_0 < 0\}.$$

Since we also know that $u_0 \in C^{1,1}_{\mathrm{loc}}(\mathbb{R}^n)$, this can be combined into

$$\Delta u_0 = f(x^0, u(x^0)+)\chi_{\{u_0>0\}} + f(x^0, u(x^0)-)\chi_{\{u_0<0\}} \quad \text{in} \quad G_0.$$

Finally, for any $x \in \mathbb{R}^n \setminus G_0$ there exists $x^{j_k} \in B_{R/r_{j_k}} \setminus G(u_{r_{j_k}})$ such that $x^{j_k} \to x$. Then the equalities $|\nabla u_{r_{j_k}}(x^{j_k})| = 0$ imply that $|\nabla u_0(x)| = 0$. □

Proof of Theorem 3.18. The proof is based on an application of the almost monotonicity formula in Theorem 2.13.

For a unit vector e and $r > 0$ define

$$\begin{aligned}\phi_e(r, u, x^0) &:= \Phi(r, (\partial_e u)^+, (\partial_e u)^-, x^0)\\ &= \frac{1}{r^4}\int_{B_r(x^0)} \frac{|\nabla(\partial_e u)^+|^2\,dx}{|x-x^0|^{n-2}} \int_{B_r(x^0)} \frac{|\nabla(\partial_e u)^-|^2\,dx}{|x-x^0|^{n-2}}.\end{aligned}$$

Now, from $W^{2,p}$ convergence $u_{x^0,r_j} \to u_0$ we have that

$$\phi_e(r, u_0, 0) = \lim_{j\to\infty} \phi_e(r, u_{x^0,r_j}, 0).$$

On the other hand, it is easy to see that we have the rescaling property

$$\phi_e(r, u_{x^0,r_j}, 0) = \phi_e(r r_j, u, x^0).$$

Further, recall that by Lemma 2.15 we have that $\Delta(\partial_e u)^\pm \geq -L$. Besides, we know that $|(\partial_e u)^\pm(x)| \leq C|x - x^0|$. Then by Theorem 2.13 we know that the limit $\phi_e(0+, u, x^0)$ exists. As a consequence, we obtain that

$$\phi_e(r, u_0, 0) = \lim_{j\to\infty} \phi_e(r r_j, u, x^0) = \phi_e(0+, u, x^0)$$

is constant for $r > 0$. Besides, observe that by Lemma 2.15 $(\partial_e u_0)^\pm$ are subharmonic, since u_0 solves an obstacle-type problem identified in Lemma 3.20, for which the constants M_1 and M_2 in $\mathbf{OT}_2$ can be taken equal to 0.

Now, from the case of equality in the ACF monotonicity formula (see Theorem 2.9), either one or the other of the following statements holds:

(i) one of the functions $(\partial_e u_0)^+$ and $(\partial_e u_0)^-$ vanishes identically in $\mathbb{R}^n$;

(ii) $(\partial_e u_0)^\pm = k_\pm (x \cdot \omega)^\pm$, $x \in \mathbb{R}^n$, for some constants $k_\pm = k_\pm(e) > 0$ and a unit vector $\omega = \omega(e)$.

To finish the proof we need to consider several possibilities.

1) Suppose $\{|\nabla u_0| = 0\} \supset \mathbb{R}^n \setminus G_0$ has a positive Lebesgue measure. Then (ii) above cannot hold for any direction e. Consequently, (i) holds for any direction e, which is equivalent to u_0 being a monotone function of one variable; see Exercise 3.5.

2) Suppose now that $|\{|\nabla u_0| = 0\}| = 0$. Then u_0 satisfies

$$\Delta u_0 = a_+ \chi_{\{u_0>0\}} + a_- \chi_{\{u_0<0\}} \quad \text{a.e. in } \mathbb{R}^n,$$

where $a_+ \geq a_-$.

2a) If $a_+ = a_- = a$ or u_0 has a sign (meaning that either $u_0 \geq 0$ or $u_0 \leq 0$ in the entire $\mathbb{R}^n$), then $\Delta u_0 = \mathit{const}$ a.e. in $\mathbb{R}^n$. Note that one needs to use that $|\{u_0 = 0\}| = 0$ for this conclusion. The latter follows from the fact that $|\nabla u_0| = 0$ a.e. on $\{u_0 = 0\}$ and the assumption that $|\{|\nabla u_0| = 0\}| = 0$. Now, by the Liouville theorem, u_0 must be a quadratic polynomial. Besides, it must be homogeneous since $u_0(0) = |\nabla u_0(0)| = 0$.

2b) Suppose now that $a_+ > a_-$ and that both $u_0^\pm$ are nonzero. Assume additionally that there exists y^0 such that $u_0(y^0) = 0$ and $\nabla u_0(y^0) \neq 0$. Let $\nu^0 = \nabla u_0(y^0)/|\nabla u_0(y^0)|$ be the direction of the gradient of u_0 at y^0. Then there exists a small neighborhood $B_\rho(y^0)$ where $\partial_{\nu^0} u_0 > 0$, such that $\{u_0 = 0\} \cap B_\rho(y^0)$ is a $C^{1,\alpha}$-surface. If $e \cdot \nu^0 \neq 0$, then $\partial_e u_0(y^0) \neq 0$ and for sufficiently small δ we have (see Exercise 3.6)

$$|\Delta \partial_e u_0|(B_\delta(y^0)) = (a_+ - a_-) \int_{\{u_0=0\}\cap B_\delta(y^0)} |e \cdot \nu| \, dH^{n-1} > 0,$$

where $\nu = \nabla u_0/|\nabla u_0|$ is the normal to the surface $u_0 = 0$. Since linear functions are harmonic, alternative (ii) cannot hold for directions e nonorthogonal to ν^0. Therefore, (i) holds for all such directions. Then, by continuity, (i) holds for all directions. As before, this implies that u_0 is a monotone function of one variable.

2c) Finally, we consider the case when $a_+ > a_-$, both $u_0^\pm$ are nonzero, and $|\nabla u_0|$ vanishes wherever u_0 vanishes. Suppose also that alternative (ii) holds for some direction e, and that we have the representation $(\partial_e u_0)^\pm = k_\pm (x \cdot \omega)^\pm$ with some $k_\pm > 0$ and $|\omega| = 1$. Since $\partial_e u_0$ does not vanish on open half-spaces $\{\pm x \cdot \omega > 0\}$, from the assumptions in this subcase it follows that u_0 does not vanish there either. Hence, u_0 does not change its sign in these half-spaces. Moreover, since both phases $u_0^\pm$ must be nontrivial, u_0 must be positive in one of the half-spaces and negative in the other one. As a consequence, u_0 must vanish on the hyperplane $\{x \cdot \omega = 0\}$. This structure of u_0 uniquely determines ω up to a sign: if e' is any other direction such that alternative (ii) holds, then the corresponding direction $\omega(e')$ must be either ω or $-\omega$. But then if $e' \cdot \omega \neq 0$, integrating from the hyperplane $\{x \cdot \omega = 0\}$ in directions $\pm e'$, we will obtain $u_0 \geq 0$ or $u_0 \leq 0$ everywhere in

$\mathbb{R}^n$, contrary to our assumption. Thus, for any direction e' nonorthogonal to ω, alternative (i) holds. By continuity, (i) holds for all directions, and we conclude as before. (A different proof for subcase 2c is given in Exercise 3.8.)

This completes the proof of the theorem. □

Theorem 3.21 (Unique type of blowups: $\mathbf{OT}_1$–$\mathbf{OT}_2$). *Let u and x^0 be as in Theorem 3.18. Then either all possible strong blowups of u at x^0 are monotone functions of one variable, or all of them are homogeneous polynomials.*

Proof. This follows immediately from the analysis of the proof of Theorem 3.18. If there exists a blowup u_0 which is a monotone function of one variable, then

$$\phi_e(0+, u, x^0) = \phi_e(r, u_0, 0) = 0,$$

for any direction e. This, in turn, implies that for any other blowup u_0'

$$\phi_e(r, u_0', 0) = \phi_e(0+, u, x^0) = 0,$$

for any direction e and $r > 0$. The latter is possible only if u_0' is a monotone function of one variable. The proof is complete. □

With the use of Theorem 3.18 we can give a full description of blowups in our model Problems **A**, **B**, and **C**.

Theorem 3.22 (Classification of blowups: Problems **A**, **B**, **C**). *Let u_0 be a blowup of a solution of Problem* **A**, **B**, *or* **C**. *Then u_0 has one of the following forms.*

– *In Problems* **A**, **B**:

- Polynomial solutions $u_0(x) = \frac{1}{2}(x \cdot Ax)$, $x \in \mathbb{R}^n$, *where A is an $n \times n$ symmetric matrix with* $\operatorname{Tr} A = 1$.
- Half-space solutions $u_0(x) = \frac{1}{2}[(x \cdot e)^+]^2$. $x \in \mathbb{R}^n$, *where e is a unit vector.*

– *In Problem* **C**:

- Polynomial solutions (positive or negative) $u_0(x) = \frac{\lambda_+}{2}(x \cdot Ax)$ *or* $u_0(x) = -\frac{\lambda_-}{2}(x \cdot Ax)$, $x \in \mathbb{R}^n$, *where A is an $n \times n$ nonnegative symmetric matrix with* $\operatorname{Tr} A = 1$.
- Half-space solutions (positive or negative) $u_0(x) = \frac{\lambda_+}{2}[(x \cdot e)^+]^2$ *or* $u_0(x) = -\frac{\lambda_-}{2}[(x \cdot e)^-]^2$, $x \in \mathbb{R}^n$, *for a unit vector e.*
- Two-half-space solutions $u_0(x) = \frac{\lambda_+}{2}[(x \cdot e)^+]^2 - \frac{\lambda_-}{2}[(x \cdot e)^-]^2$, $x \in \mathbb{R}^n$, *for a unit vector e.*

Proof. The proof consists in identifying the quadratic polynomial and one-dimensional solutions of Problems **A**, **B**, **C**. This is left to the reader as an exercise (Exercise 3.9). □

Theorem 3.23 (Unique type of blowup: Problems **A**, **B**, **C**). *Let $u \in P_R(x^0, M)$ be a solution of Problem* **A**, **B**, *or* **C**. *Then all possible blowups of u at x^0 have the same type, i.e., they fall into the same category as described in Theorem* 3.22.

Proof. The proof follows easily from Theorem 3.21 as well as Lemma 3.6 in the case of Problem **C**. □

Theorems 3.22 and 3.23 above lead to the following classification of free boundary points.

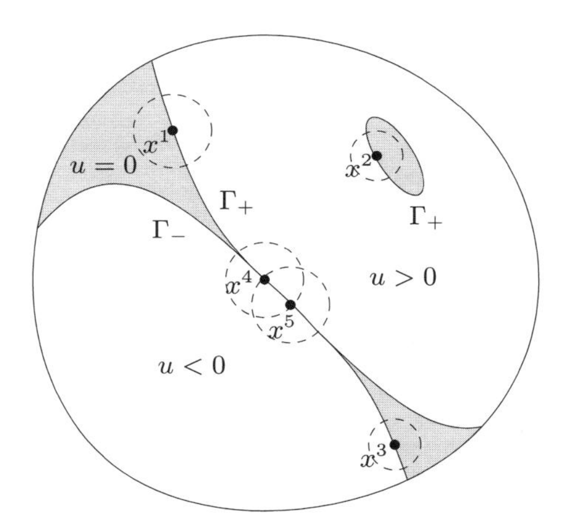

Figure 3.2. Problem **C**: Free boundary $\Gamma(u) = \Gamma_+(u) \cup \Gamma_-(u)$; x^1, x^2, x^3 are one-phase points; x^4 is a branch point ($|\nabla u(x^4)| = 0$); x^5 is a nonbranch two-phase point ($|\nabla u(x^5)| > 0$)

Definition 3.24 (Classification of free boundary points).

– In Problems **A** and **B** for $x^0 \in \Gamma(u)$ we will use the following terminology:

- x^0 is a *regular point* if one and therefore every blowup of u at x^0 is a half-space solution;
- x^0 is a *singular point* if one and therefore every blowup of u at x^0 is polynomial.

– In Problem **C**, for $x^0 \in \Gamma(u)$, we say that

- x^0 is a *positive* or *negative one-phase point* if $x^0 \in \Gamma_+(u) \setminus \Gamma_-(u)$ or $x^0 \in \Gamma_-(u) \cap \Gamma_+(u)$, respectively;

  - we further distinguish *regular* and *singular* one-phase points, similarly to Problem **A**.
- x^0 is a *two-phase point* if $x^0 \in \Gamma_+(u) \cap \Gamma_-(u)$;
  - if additionally $x^0 \in \Gamma^0(u)$, we say that x^0 is a *branch point*;
  - otherwise, if $x^0 \in \Gamma^*(u)$, we say that x^0 is a *nonbranch two-phase point.*

The classification of free boundary points in Problem **C** is illustrated in Fig. 3.2.

3.5. Weiss-type monotonicity formulas

In this section we introduce a useful tool in the study of Problems **A**, **B**, **C**: the so-called Weiss's monotonicity formula [**Wei99b**]. We are going to present it for solutions of all three model problems, even though it is most useful in Problems **A** and **C**.

Definition 3.25. For a solution $u \in P_R(x^0, M)$ and $0 < r < R$ define *Weiss's energy functional* as follows:

- In Problem **A**

$$W(r,u,x^0) := \frac{1}{r^{n+2}} \int_{B_r(x^0)} (|\nabla u|^2 + 2u)\, dx - \frac{2}{r^{n+3}} \int_{\partial B_r(x^0)} u^2 dH^{n-1}. \tag{3.10}$$

- In Problem **B**

$$W(r,u,x^0) := \frac{1}{r^{n+2}} \int_{B_r(x^0)} (|\nabla u|^2 + 2[u - u(x^0)])\, dx - \frac{2}{r^{n+3}} \int_{\partial B_r(x^0)} [u - u(x^0)]^2 dH^{n-1}. \tag{3.11}$$

- In Problem **C**

$$W(r,u,x^0) := \frac{1}{r^{n+2}} \int_{B_r(x^0)} (|\nabla u|^2 + 2\lambda_+ u^+ + 2\lambda_- u^-)\, dx - \frac{2}{r^{n+3}} \int_{\partial B_r(x^0)} u^2 dH^{n-1}. \tag{3.12}$$

We will also use the abbreviated notation $W(r,u)$ for $W(r,u,0)$.

The functional W has the following scaling property:

$$W(rs,u,x^0) = W(s,u_r,0), \tag{3.13}$$

for any $0 < r < R$, $0 < s < R/r$, where $u_r = u_{x^0,r}$ is the rescaling as in (3.9). This observation leads to a simple formula for $\frac{d}{dr}W$ and ultimately to the proof of the monotonicity formula.

To simplify the notation, we introduce the following operators:

$$\begin{aligned}\partial'_{(x^0)}v &:= (x-x^0)\cdot\nabla v(x) - 2[v(x)-v(x^0)],\\ \partial' v &:= \partial'_{(0)}v.\end{aligned}$$

Theorem 3.26 (Weiss's monotonicity formula). *Let $u \in P_R(x^0, M)$ be a solution of Problem* **A**, **B**, *or* **C**. *In the case of Problem* **B** *assume additionally that*

$$u(x^0) = \sup_{\Omega^c(u)\cap B_R(x^0)} u.$$

Then $r \mapsto W(r,u,x^0)$ is a nondecreasing absolutely continuous function for $0<r<R$ and

$$\frac{d}{dr}W(r,u,x^0) \geq \frac{2}{r^{n+4}}\int_{\partial B_r(x^0)} |\partial'_{(x^0)}u|^2 dH^{n-1}, \tag{3.14}$$

for a.e. $0<r<R$.

Moreover, the identity $W(r,u,x^0) \equiv const$ for $r_1 < r < r_2$ implies the homogeneity of $u - u(x^0)$ with respect to x^0, i.e.

$$(x-x^0)\cdot\nabla u(x) - 2[u(x)-u(x^0)] \equiv 0 \qquad \text{in } B_{r_2}(x^0)\setminus B_{r_1}(x^0). \tag{3.15}$$

Remark 3.27. In Problems **A** and **C** we actually have equality in (3.14).

Proof. We only prove the differentiation formula (3.14), as the rest of the theorem is a simple corollary.

Problem **A**. Using the scaling property (3.13) we have

$$\begin{aligned}\frac{d}{dr}W(r,u,x^0) &= \frac{d}{dr}W(1,u_r)\\ &= \int_{B_1}\frac{d}{dr}(|\nabla u_r|^2 + 2u_r)\,dx - 2\int_{\partial B_1}\frac{d}{dr}(u_r^2)dH^{n-1}.\end{aligned}$$

If we now use that

$$\begin{aligned}\frac{d}{dr}(\nabla u_r) &= \nabla\frac{du_r}{dr},\\ \frac{du_r}{dr} &= \frac{\partial' u_r}{r},\end{aligned}$$

then integrating by parts, we obtain

$$\frac{d}{dr}W(r,u,x^0) = \frac{2}{r}\int_{B_1}(-\Delta u_r + 1)\partial' u_r\,dx + \frac{2}{r}\int_{\partial B_1}(\partial_\nu u_r - 2u_r)\partial' u_r\,dH^{n-1},$$

where $\partial_\nu u_r$ is the outer normal derivative of u_r on ∂B_1. Finally, noting that $(-\Delta u_r + 1)\partial' u_r = 0$ a.e. for solutions of Problem **A** and that $\partial_\nu u_r = x \cdot \nabla u_r$ on ∂B_1, we obtain

$$\frac{d}{dr} W(r, u, x^0) = \frac{2}{r} \int_{\partial B_1} |\partial' u_r|^2 dH^{n-1},$$

which implies (3.14) after scaling.

Problem **B**. Arguing precisely as in Problem **A**, we obtain that

$$\frac{d}{dr} W(r, u, x^0) = -\frac{4}{r} \int_{B_1 \cap \Omega^c(u_r)} u_r \, dx + \frac{2}{r} \int_{\partial B_1} |\partial' u_r|^2 dH^{n-1},$$

where in the last step we use the identity

$$(-\Delta u_r + 1)\partial' u_r = -2 u_r \chi_{\Omega^c(u_r)}$$

for solutions of Problem **B**. Further, the additional assumption that $u(x^0) \geq u(x)$ for every $x \in \Omega^c(u) \cap B_R(x^0)$ implies that $u_r \chi_{\Omega^c(u_r)} \leq 0$ in $B_{R/r}$ and therefore

$$\frac{d}{dr} W(r, u, x^0) \geq \frac{2}{r} \int_{\partial B_1} |\partial' u_r|^2 dH^{n-1}.$$

Problem **C**. In this case, the proof is almost identical to the one for Problem **A** and requires only minor adjustments. (This constitutes Exercise 3.10.) □

Corollary 3.28 (Homogeneity of blowups). *Let $u \in P_R(x^0, M)$ and let $u_0(x) = \lim_{j\to\infty} u_{x^0,\lambda_j}$ be a blowup of u at x^0. Then u_0 is homogeneous of degree two, i.e.*

$$u_0(\lambda x) = \lambda^2 u_0(x), \quad x \in \mathbb{R}^n, \ \lambda > 0.$$

Of course, the statement of this corollary is contained in Theorem 3.22, since we classified all blowups at free boundary points. Nevertheless, we prefer to give a direct proof of this important fact based on Weiss's monotonicity formula, since it is extremely simple and elegant.

Proof. We may assume that the convergence $u_{x^0,\lambda_j} \to u_0$ is in $C^{1,\alpha}_{\mathrm{loc}}(\mathbb{R}^n)$. Then, we have

$$W(r, u_0) = \lim_{j\to\infty} W(r, u_{x^0,\lambda_j}) = \lim_{j\to\infty} W(\lambda_j r, u, x^0) = W(0+, u, x^0),$$

for any $r > 0$, which means that $W(r, u_0)$ is constant. Hence, using the second part of Theorem 3.26, we obtain that u_0 is homogeneous of degree 2 in $\mathbb{R}^n$. □

3.5.1. Balanced energy: Problem A. Weiss's monotonicity formula leads to the following definition. We give it only in the case of Problem **A**, since it has only a limited applicability in Problems **B** and **C**.

Definition 3.29 (Balanced energy: Problem **A**). Let $u \in P_R(x^0, M)$ be a solution of Problem **A**. Then the limit

$$\omega(x^0) := W(0+, u, x^0) = \lim_{r \to 0} W(r, u, x^0), \tag{3.16}$$

which exists by Theorem 3.26, is called the *balanced energy* of u at x^0. More generally, if u solves Problem **A** in a domain D, (3.16) defines a function $\omega : \Gamma(u) \to \mathbb{R}$.

As we saw in the proof of Corollary 3.28,

$$\omega(x^0) = W(r, u_0) \equiv W(1, u_0),$$

for any blowup u_0 of u at x^0. Thus, the balanced energy at a point coincides with the Weiss energy of any of the blowups at x^0. This has two consequences: first that the balanced energy can take only a limited number of values and second that all blowups at x^0 are of the same type.

Proposition 3.30. *Let u be a solution of Problem* **A**. *Then the balanced energy is an upper semicontinuous function of $x^0 \in \Gamma(u)$ which takes only two values. More precisely, there exist a dimensional constant $\alpha_n > 0$ such that*

$$\omega(x^0) \in \left\{\alpha_n, \frac{\alpha_n}{2}\right\},$$

for any $x^0 \in \Gamma(u)$.

Proof. The upper semicontinuity follows from the fact that

$$W(r, u, \cdot) =: \omega_r(\cdot) \searrow \omega(\cdot) \qquad \text{as } r \searrow 0,$$

and the functions $\omega_r(\cdot)$ are continuous for any $r > 0$.

The second part of the proposition follows from the direct computations of the Weiss energy for homogeneous global solutions given in the next lemma. □

Lemma 3.31. *Let $u_0 \in P_\infty(M)$ be a homogeneous global solution of Problem* **A**. *Then there exist a constant $\alpha_n > 0$ such that*

$$W(r, u_0) \equiv \begin{cases} \alpha_n/2 & \text{if } u_0 \text{ is a half-space solution,} \\ \alpha_n & \text{if } u_0 \text{ is a polynomial solution.} \end{cases}$$

Proof. Integrating by parts in the expression for $W(r, u_0)$ and using that $\Delta u_0 = 1$ in $\Omega(u_0)$, we obtain that

$$W(r,u_0) \equiv W(1,u_0) = \int_{B_1} (|\nabla u_0|^2 + 2u_0)\,dx - 2\int_{\partial B_1} u_0^2\,dH^{n-1}$$
$$= \int_{B_1} (-\Delta u_0 + 2)u_0\,dx - \int_{\partial B_1} \partial' u_0 u_0\,dH^{n-1} = \int_{B_1} u_0\,dx.$$

Thus, if we denote

$$\alpha_n = \frac{1}{2}\int_{B_1} x_n^2\,dx = \frac{1}{2n}\int_{B_1} |x|^2 dx = \frac{H^{n-1}(\partial B_1)}{2n(n+2)},$$

then for polynomial solutions $u_0(x) = \frac{1}{2}(x \cdot Ax)$ we can compute that

$$W(r, \tfrac{1}{2}(x\cdot Ax)) = \frac{1}{2}\int_{B_1} x\cdot Ax\,dx = \alpha_n \operatorname{Tr} A = \alpha_n.$$

On the other hand, for half-space solutions $u_0(x) = \frac{1}{2}[(x\cdot e)^+]^2$ we have

$$W(r, \tfrac{1}{2}[(x\cdot e)^+]^2) = \frac{1}{2}\int_{B_1} [(x\cdot e)^+]^2 dx = \frac{\alpha_n}{2}. \qquad \square$$

In fact Lemma 3.31 gives also the following classification of free boundary points in terms of balanced energy.

Proposition 3.32 (Energetic classification of free boundary points). *Let u be a solution of Problem* **A** *and $x^0 \in \Gamma(u)$. Then*

$$x^0 \text{ is regular} \iff \omega(x^0) = \frac{\alpha_n}{2},$$
$$x^0 \text{ is singular} \iff \omega(x^0) = \alpha_n.$$

Notes

Nondegeneracy in the classical obstacle problem was first proved by Caffarelli-Rivière [**CR76**]. Our version for Problem **A** is more akin to the one in Caffarelli [**Caf77**] and [**Caf80**] with a similar proof, as observed by Caffarelli-Karp-Shahgholian [**CKS00**]. Nondegeneracy in Problem **B** is due to Caffarelli-Salazar [**CS02**] and the one in Problem **C** is due to Weiss [**Wei01**].

The fact that the free boundary has a locally finite $(n-1)$-Hausdorff measure in the classical obstacle problem was proved in a small note by Caffarelli [**Caf81**]; see also [**Caf98a**]. Earlier, Brezis-Kinderlehrer [**BK74**] showed that the coincidence set has a locally finite perimeter. The local finiteness of the $(n-1)$-Hausdorff measure in Problem **B** was established by Caffarelli-Salazar [**CS02**] and in Problem **C** by Weiss [**Wei01**].

The systematic approach of considering the classes of local and global solutions, as well as rescaling and blowups, goes back to the paper of Caffarelli [**Caf80**]. The purpose of that paper was to simplify and streamline some of the earlier results. This was motivated by a similar approach in geometric measure theory.

The classification of blowups in the classical obstacle problem was known already in [**Caf80**]. The classification for Problem **A** was done in Caffarelli-Karp-Shahgholian [**CKS00**], for Problem **B** in Caffarelli-Salazar-Shahgholian [**CSS04**], and for Problem **C** in Shahgholian-Uraltseva-Weiss [**SUW04**]. Our Theorem 2.14 is an attempt to uniformize the proofs in those three problems.

Weiss's monotonicity formula for the classical obstacle problem appeared for the first time in the work of Ou [**Ou94**] and then was rediscovered by Weiss [**Wei99b**]. Since then it became one of the main tools in the study of free boundary problems of a variational nature with a homogeneous structure. Earlier, a related monotonicity formula by Spruck [**Spr83**] was used to achieve similar results. Weiss's monotonicity formula serves as the basis for the energy approach (as opposed to the geometric approach of Caffarelli) for the study of the free boundary. There is also a parabolic version of Weiss's formula found by Weiss [**Wei99a**], which plays an equally important role in parabolic problems.

Exercises

3.1. For Problems **A**, **B**, prove that in any neighborhood of $x^0 \in \Gamma(u)$ there are points x such that $u(x) > u(x^0)$.

Hint: Use subharmonicity of u.

3.2. Complete the proof of Lemma 3.6.

3.3. Complete the proof of Corollary 3.7.

3.4. Prove that if E is bounded and for every ball $B = B_r(x)$ centered at $x \in E$ there exists a ball $B' = B_{\delta r}(y)$ such that $B' \subset B \setminus E$, then E is $c_n\delta$-porous. In other words, for a bounded set, it is enough to verify the porosity condition only for balls centered at the points of this set.

3.5. Prove that if $u \in C^1(\mathbb{R}^n)$ and if every derivative $\partial_e u$ has a sign in $\mathbb{R}^n$, then u is a monotone function of one variable, i.e. there exist a monotone $\phi \in C^1(\mathbb{R})$ and a direction e such that

$$u(x) = \phi(x \cdot e), \quad x \in \mathbb{R}^n.$$

Proof. Observe that if u is not identically constant, then there is a point $z \in D$ where $\nabla u(z)$ does not vanish. Let $e_0 = \nabla u(z)/|\nabla u(z)|$. Then $\partial_e u(z) > 0$ for all directions e with $e \cdot e_0 > 0$, and $\partial_e u(z) < 0$ if $e \cdot e_0 < 0$. By the assumption these inequalities will hold everywhere in $\mathbb{R}^n$. In particular, by continuity we will obtain that $\partial_e u = 0$ for all directions e, orthogonal to e_0. □

3.6. Let $u \in C^{1,1}(D)$ satisfy the equation $\Delta u = a_+\chi_{\{u>0\}} + a_-\chi_{\{u<0\}}$ in D for some constants $a_\pm$. Suppose moreover that $\nabla u \neq 0$ on the level set $\gamma = \{u = 0\}$. Then $\gamma = \{u = 0\}$ is locally a $C^{1,1}$ graph. Prove that any directional derivative $v = \partial_e u$ satisfies the equation

$$\Delta v = (a_+ - a_-)(e \cdot \nu) H^{n-1} \lfloor \gamma \quad \text{in } D,$$

where $\nu = \nabla u/|\nabla u|$ is the normal to the surface γ.

Similarly, find an equation satisfied by $|u|$.

3.7. Let D be an open subset of $\mathbb{R}^n$ and $\eta \in W_0^{1,2}(D) \cap W^{1,\infty}(D)$. If F is a closed subset of D and $\eta = 0$ on F, then $\eta \in W_0^{1,2}(D \setminus F)$.

Hint: Let $d(x) = \operatorname{dist}(x, F)$ in D. Then $d \in W^{1,\infty}(D)$ and $|\nabla d| \leq 1$ a.e. in D. Then show that the functions

$$\eta_\varepsilon(x) = \eta(x)\zeta_\varepsilon(d(x))$$

with

$$\zeta_\varepsilon(s) = \begin{cases} 0, & s \leq \varepsilon, \\ \frac{1}{\varepsilon}(s - \varepsilon), & \varepsilon < s < 2\varepsilon, \\ 1, & s \geq 2\varepsilon, \end{cases}$$

converge to η as $\varepsilon \to 0+$ in $W^{1,2}(D \setminus F)$.

3.8. Let u_0 be as in subcase 2c of the proof of Theorem 3.18. Show that $v = u_0^\pm$ satisfies $\Delta v = a\chi_{\{v>0\}}$ in $\mathbb{R}^n$ with $a = \pm a_\pm$. Using that global solutions of the classical obstacle problem are convex (see Theorem 5.1), conclude that u_0 depends only on one direction. This gives an alternative proof for the subcase 2c in Theorem 3.18.

3.9. Complete the proof of Theorems 3.22 and 3.23.

3.10. Complete the proof of Theorem 3.26 for Problem **C**.

3.11. Prove that if $x^0 \in \Gamma(u)$ in Problem **A** and for some $\rho = \rho(x_0) > 0$ we have

$$x^0 - (\mathcal{C}_\delta \cap B_\rho) \subset \{u = 0\},$$

where $\mathcal{C}_\delta = \{x \in \mathbb{R}^n : x_n \geq \delta|x'|\}$, then x^0 is a regular point. Here $\delta > 0$ is not necessarily small. Similarly, for Problem **B**: if $x^0 - (\mathcal{C}_\delta \cap B_\rho) \subset \{u = u(x^0)\}$, then x^0 is regular.

Chapter 4

Regularity of the free boundary: first results

In this chapter, we show how to use a geometric property of directional monotonicity of global solutions to derive a similar property for local solutions. We will then use that to prove some regularity results for the free boundary. More specifically: in §4.1 we prove the Lipschitz and C^1 regularity of the free boundary near regular points in Problem **A**, and in §4.2 we study the structure of the "patches" near regular points in Problem **B**. In §4.3 we prove the C^2 regularity of the solution up to the boundary near regular points in Problems **A** and **B**. This then gives enough regularity to apply the partial hodograph-Legendre transform to conclude the real analyticity of the free boundary. In §4.4 we show that $\Gamma_\pm$ are graphs of C^1 functions near branch points in Problem **C**. We conclude the chapter by showing the real analyticity of Γ^* in Problem **C** (§4.5).

In this chapter, unless otherwise stated, all the constants are strongly dependent on the solutions. However, as we will see in Chapter 6, they can be made uniform under certain geometric conditions on the solutions.

4.1. Problem A: C^1 regularity of the free boundary near regular points

In this section, we study the free boundary in Problem **A** near so-called regular points in the sense of Definition 3.24. Recall that this means that the rescalings

$$u_r(x) = u_{x^0,r}(x) = \frac{u(x^0 + rx)}{r^2}$$

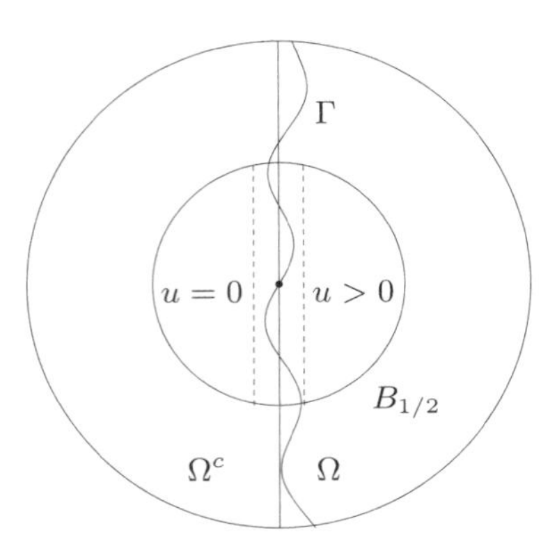

Figure 4.1. Flatness of $\Gamma(u)$

converge, over a subsequence $r = r_j \to 0$, to a half-space solution

$$u_0(x) = \frac{1}{2}[(x \cdot e)^+]^2,$$

for some unit vector e. Further, by Proposition 3.32, the free boundary point x^0 is regular iff $\omega(x^0) = W(0+, u, x^0) = \frac{\alpha_n}{2}$. An immediate corollary from the upper semicontinuity of the balanced energy (see Proposition 3.30) is that the set of regular points is a relatively open subset of $\Gamma(u)$. Our purpose in this section is to show that this set is locally a graph of a C^1 function.

The careful study of the convergence of rescalings u_{r_j} to a half-space solution u_0 will ultimately imply the regularity of $\Gamma(u)$. We start with a lemma that the closeness to a half-space solution in the L^∞-norm implies a certain flatness property for $\Gamma(u)$ (see Fig. 4.1).

Lemma 4.1. *Let $u \in P_1(M)$ and let a half-space solution $u_0(x) = \frac{1}{2}(x_n^+)^2$ be such that*

$$\|u - u_0\|_{L^\infty(B_1)} \leq \varepsilon.$$

Then

$$\begin{aligned} u > 0 \quad & \text{in } \{x_n > \sqrt{2\varepsilon}\} \cap B_1, \\ u = 0 \quad & \text{in } \{x_n \leq -2\sqrt{n\varepsilon}\} \cap B_{1/2}. \end{aligned}$$

In particular,

$$\Gamma(u) \cap B_{1/2} \subset \{|x_n| \leq 2\sqrt{n\varepsilon}\}.$$

Proof. The first statement is evident, so we will prove only the second one. Take any $x^0 \in \Omega(u) \cap B_{1/2} \cap \{x_n < 0\}$ and let $r := -x_n^0 > 0$. Then by the nondegeneracy we have

$$\sup_{B_r(x^0)} u \geq u(x^0) + c\,r^2, \quad c = \frac{1}{2n}.$$

Now, since $B_r(x^0) \subset \{x_n < 0\}$, $u_0 = 0$ in $B_r(x^0)$ and consequently $|u| \leq \varepsilon$ there. Therefore, $c\,r^2 \leq 2\varepsilon$, or, equivalently, $r \leq \sqrt{2\varepsilon/c}$, which implies that $\Omega(u) \cap B_{1/2} \subset \{x_n > -\sqrt{2\varepsilon/c}\}$. □

As an immediate corollary we obtain the following vanishing flatness property of the free boundary at regular points.

Proposition 4.2 (Flatness of $\Gamma(u)$). *Let $u \in P_1(M)$ be a solution of Problem* **A** *such that 0 is a regular point. Then for any $\sigma > 0$ there exists $r_\sigma = r_\sigma(u) > 0$ such that $\Gamma(u)$ is σ-flat in B_r for any $r < r_\sigma$ in the sense that*

$$\Gamma(u) \cap B_r \subset \{|x \cdot e| < \sigma r\}$$

for some direction $e = e_{r,\sigma}$.

Proof. Suppose the statement fails for some $\sigma > 0$. Then there exists a sequence $r_j \to 0$ such that $\Gamma(u)$ is not σ-flat in $B_{r_j/2}$, which is equivalent to saying that $\Gamma(u_{r_j})$ is not σ-flat in $B_{1/2}$. Without loss of generality we may assume that $u_{r_j} \to u_0$ in $C^{1,\alpha}_{\mathrm{loc}}(\mathbb{R}^n)$, where $u_0(x) = \frac{1}{2}(x_n^+)^2$. Then for $j \geq j_\varepsilon$ we will have that $\|u_{r_j} - u_0\|_{L^\infty(B_1)} \leq \varepsilon$ and therefore by Lemma 4.1

$$\Gamma(u_{r_j}) \cap B_{1/2} \subset \{|x_n| < c_n\sqrt{\varepsilon}\},$$

which will make $\Gamma(u_{r_j})$ σ-flat in $B_{1/2}$ if $c_n\sqrt{\varepsilon} < \frac{1}{2}\sigma$, a contradiction. □

Next, we show that the closeness of rescalings u_{r_j} to u_0 in the C^1-norm implies the Lipschitz regularity of $\Gamma(u)$. To be more precise, we show that the directional derivatives $\partial_e u$ are nonnegative near a regular point for a certain cone of directions. This will follow from the fact that the half-space solution $u_0(x) = \frac{1}{2}(x_n^+)^2$ is monotone increasing in any direction $e \in \mathcal{C}_\delta \cap \partial B_1$, where

$$\mathcal{C}_\delta := \{x \in \mathbb{R}^n : x_n > \delta|x'|\}, \quad x' = (x_1, \ldots, x_{n-1}).$$

In fact, we will need to use the following stronger version of the above statement:

$$\delta^{-1}\partial_e u_0 - u_0 \geq 0 \quad \text{in } B_1 \quad \text{for any } e \in \mathcal{C}_\delta \cap \partial B_1,$$

if $\delta \in (0, 1]$. As we show in the following lemma, this property is stable with respect to C^1 approximations. This is one of the key ideas that will appear in modified forms in the other problems.

Lemma 4.3. *Let u be a solution of Problem* **A** *in B_1 and suppose that for a certain direction e and a constant $C > 0$, we have*

$$C\partial_e u - u \geq -\varepsilon_0 \quad \text{in } B_1, \tag{4.1}$$

with $\varepsilon_0 < 1/8n$. Then

$$C\partial_e u - u \geq 0 \quad \text{in } B_{1/2}. \tag{4.2}$$

Proof. Suppose the conclusion of the lemma fails and let $y \in B_{1/2} \cap \Omega$ be such that $C\partial_e u(y) - u(y) < 0$. Consider then the auxiliary function

$$w(x) = C\partial_e u(x) - u(x) + \frac{1}{2n}|x-y|^2.$$

It is easy to see that w is harmonic in $\Omega \cap B_{1/2}(y)$, $w(y) < 0$, and that $w \geq 0$ on $\partial\Omega$. Hence by the minimum principle, w has a negative infimum on $\partial B_{1/2}(y) \cap \Omega$, i.e.

$$\inf_{\partial B_{1/2}(y)\cap\Omega} w < 0.$$

This can be rewritten as

$$\inf_{\partial B_{1/2}(y)\cap\Omega} (C\partial_e u - u) < -\frac{1}{8n},$$

which contradicts (4.1), since $\varepsilon_0 < 1/8n$. □

In the next lemma we show how the C^1 approximations by half-space solutions imply the directional monotonicity.

Lemma 4.4. *Let $u \in P_1(M)$, $u_0(x) = \frac{1}{2}(x_n^+)^2$ and suppose that*

$$\sup_{B_1} |u - u_0| \leq \varepsilon, \quad \sup_{B_1} |\nabla u - \nabla u_0| \leq \varepsilon, \tag{4.3}$$

for some $\varepsilon > 0$. Then

(i) $\quad \varepsilon \leq \dfrac{1}{32n} \quad \Rightarrow \quad u \geq 0 \quad$ *in $B_{1/2}$,*

(ii) $\quad \varepsilon \leq \dfrac{\delta}{32n}, \quad \delta \in (0,1] \quad \Rightarrow \quad \partial_e u \geq 0 \quad$ *in $B_{1/2}$ for any $e \in \mathcal{C}_\delta$.*

Proof. Recall that

$$\delta^{-1}\partial_e u_0 - u_0 \geq 0 \quad \text{in } B_1 \quad \text{for any } e \in \mathcal{C}_\delta \cap \partial B_1.$$

Using the closeness condition (4.3) with $\varepsilon \leq \delta/32n$, we obtain that

$$\delta^{-1}\partial_e u - u \geq -2\varepsilon\delta^{-1} \geq -\frac{1}{16n} \quad \text{in } B_1.$$

Hence, from Lemma 4.3 we deduce that

$$\delta^{-1}\partial_e u - u \geq 0 \quad \text{in } B_{1/2} \quad \text{for any } e \in \mathcal{C}_\delta \cap \partial B_1. \tag{4.4}$$

Further, by Lemma 4.1 we have that

$$u = 0 \quad \text{in } \left\{ x_n \leq -\frac{1}{2\sqrt{2}} \right\} \cap B_{1/2}.$$

In particular, taking $\delta = 1$, multiplying (4.4) by $\exp(e \cdot x)$ and integrating, we obtain (i). To finish the proof, we notice that (ii) follows from (4.4) and (i). □

The next result is essentially a rescaled version of Lemma 4.4.

Lemma 4.5 (Directional monotonicity). *Let $u \in P_1(M)$ be a solution of Problem **A** such that 0 is a regular point and $u_0 = \frac{1}{2}(x_n^+)^2$ is a blowup at 0. Then for any $\delta \in (0,1]$ there exists $r_\delta = r_\delta(u) > 0$ such that*

(i) $$u \geq 0 \quad \text{in } B_{r_1},$$

(ii) $$\partial_e u \geq 0 \quad \text{in } B_{r_\delta} \quad \text{for any } e \in \mathcal{C}_\delta.$$

Proof. The proof is left as an exercise to the reader. □

The lemma above has an immediate corollary, which we would like to point out.

Proposition 4.6 (Uniqueness of blowup). *Let $u \in P_1(M)$ be a solution of Problem **A** such that 0 is a regular point. Then the blowup at 0 is unique. That is, after a possible rotation of coordinate axes,*

$$u_r(x) \to u_0(x) = \frac{1}{2}(x_n^+)^2 \quad \text{as } r \to 0.$$

Proof. Without loss of generality we may assume that $u_{r_j} \to u_0 = \frac{1}{2}(x_n^+)^2$, over some sequence $r_j \to 0$. Suppose now that $u_{r'_j} \to u'_0 = \frac{1}{2}[(x \cdot e'_0)^+]^2$, over another sequence $r'_j \to 0$. Note that as soon as $r'_j < r_\delta$ for r_δ as in Lemma 4.5, we have

$$\partial_e u_{r'_j} \geq 0 \quad \text{for any } e \in \mathcal{C}_\delta,$$

and therefore passing to the limit we will obtain that

$$\partial_e u'_0 \geq 0 \quad \text{for any } e \in \mathcal{C}_\delta.$$

This is equivalent to saying that $e'_0 \cdot e \geq 0$ for any $e \in \mathcal{C}_\delta$. Letting $\delta \to 0$, we see that e'_0 necessarily coincides with e_n and therefore $u'_0 = u_0$. □

Lemma 4.7. *Let u and u_0 be as in Lemma 4.4 and assume that (4.3) holds with $\varepsilon \leq \delta/32n$, $\delta \in (0,1]$. Then for any $z \in \Gamma(u) \cap B_{1/2}$,*

$$u > 0 \quad \text{in } (z + \mathcal{C}_\delta) \cap B_{1/2},$$
$$u = 0 \quad \text{in } (z - \mathcal{C}_\delta) \cap B_{1/2}.$$

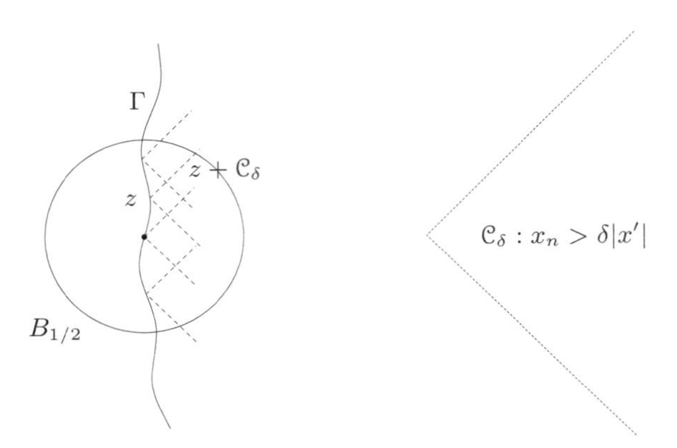

Figure 4.2. Cone of monotonicity $\mathcal{C}_\delta$ and Lipschitz regularity of the free boundary

In particular, $\Gamma(u) \cap B_{1/2}$ *is a Lipschitz graph* $x_n = f(x')$ *with the Lipschitz constant of* f *not exceeding* δ (see Fig. 4.2).

Proof. First let $z \in B_{1/2}$ be such that $u(z) = 0$. From the monotonicity of u in any direction $e \in \mathcal{C}_\delta$ in $B_{1/2}$ we immediately get

$$u \geq 0 \quad \text{in } (z + \mathcal{C}_\delta) \cap B_{1/2},$$
$$u \leq 0 \quad \text{in } (z - \mathcal{C}_\delta) \cap B_{1/2}.$$

On the other hand $u \geq 0$ in $B_{1/2}$ and therefore $u = 0$ in $(z - \mathcal{C}_\delta) \cap B_{1/2}$. Now assume that $z \in \Gamma(u) \cap B_{1/2}$ and show that this implies the strict positivity of u in $(z + \mathcal{C}_\delta) \cap B_{1/2}$. Indeed, otherwise there exists $y \in (z + \mathcal{C}_\delta) \cap B_{1/2}$ with $u(y) = 0$, and from the argument above, $u = 0$ in $(y - \mathcal{C}_\delta) \cap B_{1/2}$. The latter set, however, is a neighborhood of z, which contradicts the fact that $z \in \Gamma(u)$. Hence $u > 0$ in $(z + \mathcal{C}_\delta) \cap B_{1/2}$.

For the last statement of the lemma, see Exercise 4.1. □

Proposition 4.8 (Lipschitz regularity of $\Gamma(u)$). *Let* $u \in P_1(M)$ *be a solution of Problem* **A** *such that* 0 *is a regular point and* $u_0 = \frac{1}{2}(x_n^+)^2$ *is a blowup at* 0. *Then there exist* $\rho = \rho(u) > 0$ *and a Lipschitz function* $f : B'_\rho \to \mathbb{R}$ *such that*

$$\Omega(u) \cap B_\rho = \{x \in B_\rho : x_n > f(x')\},$$
$$\Gamma(u) \cap B_\rho = \{x \in B_\rho : x_n = f(x')\}.$$

Moreover, if r_δ, $\delta \in (0, 1]$, *is as in Lemma 4.5, then* $|\nabla_{x'} f| \leq \delta$ *a.e. on* $B'_{r_\delta/2}$.

Proof. The proof is a simple consequence of the lemmas above; see Exercise 4.2. □

In fact one extra step gives us a rather short proof of C^1-regularity of the free boundary in a neighborhood of a regular point. But first we need the following fact.

Lemma 4.9. *Let $u \in P_1(M)$ be a solution of Problem **A** such that 0 is a regular point. Then there exists $\rho > 0$ such that all points in $\Gamma(u) \cap B_\rho$ are also regular.*

The statement of the lemma follows easily from the upper semicontinuity of the balanced energy; see Proposition 3.30. However, here we prefer to give a proof that does not use Weiss's monotonicity, and that can be later modified for Problem **B**.

Proof. By Lemma 4.5, any blowup u_0 at a point $y^0 \in \Gamma(u) \cap B_{r_1}$ must satisfy $\partial_e u_0 \geq 0$ in $\mathbb{R}^n$ for any direction $e \in \mathcal{C}_1$. Then u_0 must be a half-space solution since no homogeneous quadratic polynomial can be monotone in a cone of directions $\mathcal{C}_1$. Thus, the lemma holds with $\rho = r_1$. □

Theorem 4.10. *Let $u \in P_1(M)$ be a solution of Problem **A** such that 0 is a regular point. Then there exists $\rho > 0$ such that $\Gamma(u) \cap B_\rho$ is a C^1 graph.*

Proof. We may assume that $u_0 = \frac{1}{2}(x_n^+)^2$ is the blowup at the origin. Then by Proposition 4.8, $\Gamma(u) \cap B_\rho$ is a Lipschitz graph $x_n = f(x')$, with $f(0) = 0$ and $|f(x')| \leq \delta |x'|$ in $B'_{r_\delta/2}$. Since we can choose $\delta > 0$ arbitrarily small, we immediately obtain the existence of a tangent plane to $\Gamma(u)$ at the origin with e_n being the normal vector; see Fig. 4.3.

Now, by Lemma 4.9, every free boundary point $z \in \Gamma(u) \cap B_\rho$ is regular and therefore $\Gamma(u)$ has a tangent plane at all these points. Let ν_z denote the unit normal vector at $z \in \Gamma(u)$ pointing into $\Omega(u)$. Then Lemma 4.5 implies that $\nu_z \cdot e \geq 0$ for any $e \in \mathcal{C}_\delta$ if $z \in \Gamma(u) \cap B_{r_\delta}$. This means that ν_z is from the dual of the cone $\mathcal{C}_\delta$, that is, $\nu_z \in \mathcal{C}_{1/\delta}$. In particular, we obtain

$$|\nu_z - e_n| \leq C\delta, \quad z \in \Gamma(u) \cap B_{r_\delta}.$$

The latter means that $\Gamma(u)$ is C^1 at the origin. The same is evidently true for any $z \in \Gamma(u) \cap B_\rho$. This completes the proof. □

Deeper investigation, including the uniform estimates of C^1 modulus of continuity for a class of solutions $u \in P_1(M)$, will be done in Chapter 6.

4.2. Problem B: the local structure of the patches

In this section, we study the free boundary near regular points in Problem **B**. Recall that the classification of free boundary points is given in Definition 3.24.

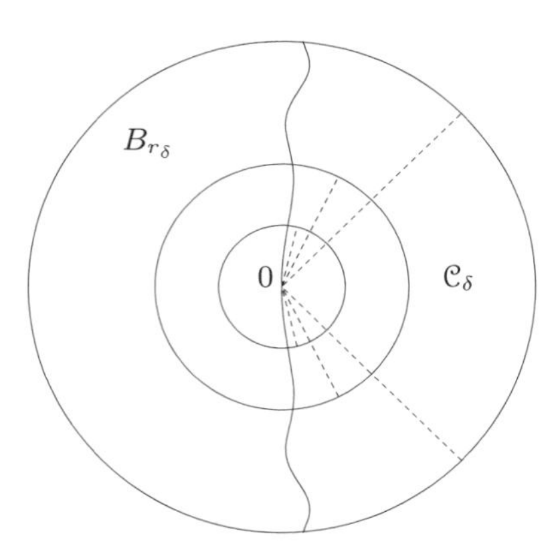

Figure 4.3. C^1 regularity of $\Gamma(u)$

The main result of this section is the following theorem. Basically, it says that near regular points, solutions of Problem **B** actually solve Problem **A**, after a possible subtraction of a constant. Thus, we can further apply the results from the previous section.

Theorem 4.11. *Let $u \in P_1(M)$ and suppose that 0 is a regular point. Then, there exists $\rho = \rho(u) > 0$ such that*

$$\Omega^c \cap B_\rho \subset \{u = u(0)\}.$$

Combining this with Theorem 4.10, we arrive at the following corollary.

Corollary 4.12. *Let u be as in Theorem 4.11 above. Then there exists $\rho = \rho(u) > 0$ such that $\Gamma(u) \cap B_\rho$ is a C^1 graph.*

The proof of Theorem 4.11 is also based on the idea of directional monotonicity, similar to that for Problem **A**, which we discussed in the previous section. As before, we need a statement that is stronger than the directional monotonicity and is stable under the blowup. For Problem **B**, it turns out to be

$$C\partial_e u - |\nabla u|^2 \geq 0,$$

for the reasons that we will see in the next lemmas.

Lemma 4.13. *Let $u \in P_1(M)$ and suppose that*

$$C\partial_e u - |\nabla u|^2 \geq -\varepsilon_0 \quad \text{in } B_1,$$

where $\varepsilon_0 < 1/4n^2$. Then

$$C\partial_e u - |\nabla u|^2 \geq 0 \quad \text{in } B_{1/2}.$$

Proof. We argue similarly to the proof of Lemma 4.3 and use that

$$\Delta(|\nabla u|^2) = 2|D^2u|^2 + 2\nabla u \nabla(\Delta u) \geq \frac{2}{n}$$

in $\Omega(u)$ instead of $\Delta u = 1$. We leave the details to the reader (Exercise 4.3). □

Lemma 4.14. *Let $u \in P_1(M)$ with 0 a regular point and suppose that $u_0(x) = \frac{1}{2}(x_n^+)^2$ is a blowup of u at 0. Then, for any $\delta \in (0,1]$ there is $r_\delta > 0$ such that*

$$(4\delta^{-1} r_\delta)\partial_e u - |\nabla u|^2 \geq 0 \quad \textit{in } B_{r_\delta} \quad \textit{for any } e \in \mathcal{C}_\delta \cap \partial B_1.$$

In particular, $\partial_e u \geq 0$ in B_{r_δ} for any $e \in \mathcal{C}_\delta$.

Proof. For the global solution u_0 we have that

$$\partial_e u_0 = x_n^+(e \cdot e_n) \quad \text{and} \quad |\nabla u_0|^2 = (x_n^+)^2.$$

Hence for $e \in \mathcal{C}_\delta$ the following stronger version of the directional monotonicity is satisfied:

$$2\delta^{-1}\partial_e u_0 - |\nabla u_0|^2 \geq 0 \quad \text{in } B_1.$$

Fix now a sequence $r_k \to 0$ such that

$$u_0 = \lim_{k\to\infty} u_{r_k}.$$

By uniform convergence, for any $\varepsilon_0 > 0$, there is k_0 such that for all $k \geq k_0$, and $e \in \mathcal{C}_\delta$ we have

$$2\delta^{-1}\partial_e u_{r_k} - |\nabla u_{r_k}|^2 \geq -\varepsilon_0 \quad \text{in } B_1.$$

Now take $\varepsilon_0 < 1/4n^2$. Then by Lemma 4.13

$$2\delta^{-1}\partial_e u_{r_k} - |\nabla u_{r_k}|^2 \geq 0 \quad \text{in } B_{1/2},$$

and we complete the proof by taking $r_\delta = \frac{r_{k_0}}{2}$. □

As a direct corollary we obtain the following uniqueness of blowups.

Proposition 4.15 (Uniqueness of blowup). *Let $u \in P_1(M)$ be a solution of Problem* **B** *such that 0 is a regular point. Then the blowup at 0 is unique. That is, after a suitable rotation of coordinate axes,*

$$u_r(x) \to u_0(x) = \frac{1}{2}(x_n^+)^2 \quad \textit{as } r \to 0.$$

Proof. The proof is obtained by a similar reasoning as that of Proposition 4.6. □

Repeating the arguments in Lemma 4.9 with $\rho = r_1$ as in Lemma 4.14, we obtain the following.

Lemma 4.16. *Let $u \in P_1(M)$ be a solution of Problem* **B** *such that* 0 *is a regular point. Then there exists $\rho = \rho(u) > 0$ such that all points in $\Gamma(u) \cap B_\rho$ are regular.* □

We next show that Ω^c contains a cone centered at a regular point, which becomes larger as we approach the point.

Lemma 4.17. *Under the assumptions of Lemma* 4.14 *with $\delta \in (0, 1/2]$ there is $r'_\delta > 0$ such that*

$$-(\mathcal{C}_{2\delta} \cap B_{r'_\delta}) \subset \Omega^c.$$

Proof. Suppose there is a sequence $\{x^j\} \subset -\mathcal{C}_{2\delta} \cap B_1 \cap \Omega$ such that $\rho_j = |x^j| \to 0$. Fix a constant $\tau > 0$, such that

$$B_{\tau|x|}(x) \subset -\mathcal{C}_\delta, \quad \text{for any } x \in -\mathcal{C}_{2\delta}.$$

By the nondegeneracy of u in Ω, there exist $y^j \in B_{\tau\rho_j}(x^j)$ such that

$$u_{\rho_j}({\rho_j}^{-1}y^j) - u_{\rho_j}({\rho_j}^{-1}x^j) \geq c_0\tau^2, \tag{4.5}$$

for some constant $c_0 > 0$, independent of j.

Now, from the assumptions on u and Proposition 4.15 we have

$$\lim_{r\to 0} u_r(x) = \frac{1}{2}(x_n^+)^2, \quad \text{for all } x \in \mathbb{R}^n.$$

Moreover, the convergence is uniform (even $C^{1,\alpha}$) on compact subsets of $\mathbb{R}^n$. To reach a contradiction, now observe that $\{{\rho_j}^{-1}y^j\}$ and $\{{\rho_j}^{-1}x^j\}$ are two bounded sequences contained in $-\mathcal{C}_\delta \cap (B_{3/2} \setminus B_{1/2}) \Subset \{x_n < 0\}$ and therefore

$$\lim_{j\to\infty} u_{\rho_j}({\rho_j}^{-1}y^j) = 0, \quad \lim_{j\to\infty} u_{\rho_j}({\rho_j}^{-1}x^j) = 0.$$

This contradicts (4.5) and completes the proof of the lemma. □

We are now ready to prove the main result of this section.

Proof of Theorem 4.11. Without loss of generality we may assume that $u_0(x) = \frac{1}{2}(x_n^+)^2$ is a blowup of u at 0.

Fix some $0 < \delta < 1/2$ and let $\rho = \frac{2}{3}\min\{r_\delta, r'_\delta\}$, with r_δ as in Lemma 4.14 and r'_δ as in Lemma 4.17.

We claim that any point $x \in \Omega^c \cap B_\rho$ can be joined to $-\mathcal{C}_{2\delta} \cap B_\rho$ by a line segment parallel to e_n and contained in Ω^c. Indeed, if the claim is not true, then we can find two points, $x^1 \in \Omega \cap B_\rho$ and $x^2 \in \Omega^c \cap B_\rho$, such that

$$x^2 - x^1 = |x^2 - x^1|e_n.$$

Now, take a small ball $B_\varepsilon(x^1) \subset \Omega$ and denote by $\mathcal{C}$ the cone generated by x^2 and $B_\varepsilon(x^1)$. Move $B_\varepsilon(x^1)$ from x^1 to x^2 along the axis of $\mathcal{C}$, reducing its radius to fit in $\mathcal{C}$, until we touch Ω^c for the first time. Let ζ^0 be a point of contact (which may coincide with x^2 as well). It is easy to see that there exist $\rho' > 0$ and $\delta' > 0$ such that

$$\zeta^0 - (\mathcal{C}_{\delta'} \cap B_{\rho'}) \subset B_\rho \cap \Omega. \tag{4.6}$$

(Note that in the case when $\zeta^0 = x^2$, as $\mathcal{C}_{\delta'}$ one can take the cone $\mathcal{C}$ spanned by x^2 and $B_\varepsilon(x^1)$.) Since by Lemma 4.14, $\partial_e u \geq 0$ in B_ρ, for all $e \in \mathcal{C}_\delta$, it follows that

$$\zeta^0 - (\mathcal{C}_\delta \cap B_{\rho'}) \subset \{u \leq u(\zeta^0)\}$$

and

$$\zeta^0 + (\mathcal{C}_\delta \cap B_{\rho'}) \subset \{u \geq u(\zeta^0)\}.$$

Now let $\hat{u}_0$ be a blowup of u at ζ^0. Then $\hat{u}_0$ cannot be a polynomial solution, since $\hat{u}_0 \leq 0$ in $-\mathcal{C}_\delta$ and $\hat{u}_0 \geq 0$ in $\mathcal{C}_\delta$ and homogeneous polynomials of degree two are even. Therefore, $\hat{u}_0$ should be a half-space solution, and hence nonnegative. In particular, $\hat{u}_0$ must vanish on $-\mathcal{C}_\delta$. On the other hand, from (4.6) it follows that $\hat{u}_0$ must satisfy $\Delta u = 1$ in $-\mathcal{C}_{\delta'}$, and since the intersection $\mathcal{C}_\delta \cap \mathcal{C}_{\delta'}$ is nonempty, we arrive at a contradiction. This proves the claim as well as the theorem since $u \equiv u(0)$ in $-\mathcal{C}_{2\delta} \cap B_\rho$ due to Lemma 4.17. □

4.3. Problems A and B: higher regularity of the free boundary

In this short section we give yet another application of the monotonicity in the cone of directions, to improve on the uniqueness of blowups near regular points and establish the continuity of the second derivatives of u in $\Omega(u)$ up to regular points on $\Gamma(u)$. This gives enough regularity to apply the partial hodograph-Legendre transform (discussed in detail in §6.4.2) to establish the real analyticity of the free boundary.

Proposition 4.18. *Let $u \in P_1(M)$ be a solution of Problem* **A** *or* **B** *with 0 a regular point, and $u_0(x) = \frac{1}{2}(x_n^+)^2$ a blowup of u at 0. Then*

$$\lim_{\substack{x \in \Omega(u) \\ x \to 0}} \partial_{x_i x_j} u(x) = \delta_{in}\delta_{jn} = \partial_{x_i x_j}(\tfrac{1}{2}x_n^2).$$

Proof. Since 0 is a regular point, in view of Theorem 4.11 we may assume that we are in the case of Problem **A**. Moreover, by Lemma 4.3 we may assume that $u \geq 0$ in B_1.

Let $x^j \in \Omega(u)$ be a sequence converging to 0, $d_j = \operatorname{dist}(x^j, \Gamma(u))$ and $y^j \in \Gamma(u) \cap \partial B_{d_j}(x^j)$. Also let $\xi^j = (x^j - y^j)/d_j$. Consider then the rescalings of u at y^j,

$$v_j(x) = u_{y^j,d_j}(x) = \frac{u(y^j + d_j x)}{d_j^2}.$$

Note that by definition we have

$$D^2 v_j(\xi^j) = D^2 u(x^j),$$

so we want to show that

$$D^2 v_j(\xi^j) \to D^2(\tfrac{1}{2}x_n^2).$$

Next note that we have uniform $C^{1,1}$ estimates locally in $\mathbb{R}^n$ for the family $\{v_j\}$ and thus may assume that $v_j \to v_0 \in P_\infty(M)$ in $C^{1,\alpha}_{\mathrm{loc}}(\mathbb{R}^n)$. Additionally we may assume that $\xi^j \to \xi^0$. Since also the functions v_j satisfy $\Delta v_j = 1$ in $B_1(\xi^j)$, we have that $v_j \to v_0$ in $C^2(B_{1/2}(\xi^0))$ and in particular that

$$D^2 v_j(\xi^j) \to D^2 v_0(\xi^0).$$

Thus, we will be done once we show that $v_0(x) = \frac{1}{2}(x_n^+)^2$ and $\xi^0 = e_n$. In order to prove that claim recall that by Lemma 4.5 for any $\delta > 0$ there exists $r_\delta > 0$ such that $\partial_e u \geq 0$ in B_{r_δ} for any direction $e \in \mathcal{C}_\delta$. This immediately implies that for any $R > 0$ we have $\partial_e v_j \geq 0$ in B_R, provided j is sufficiently large. Passing to the limit, we obtain therefore that

$$\partial_e v_0 \geq 0 \quad \text{in } \mathbb{R}^n,$$

for any direction e with $e \cdot e_n > 0$. This implies that $v(x) = \phi(x_n)$ for a monotone increasing function of x_n. Since 0 is a free boundary point, the only such function is $\frac{1}{2}(x_n^+)^2$, as claimed. Finally, note that ξ^0 is such that $|\xi^0| = 1$ and $\Delta v_0 = 1$ in $B_1(\xi^0)$. The only such point for $v_0(x) = \frac{1}{2}(x_n^+)^2$ is $\xi^0 = e_n$. □

Remark 4.19. An equivalent formulation of Proposition 4.18 is as follows. If 0 is a regular point, then the blowup of u over any sequence $y^j \in \Gamma(u)$, $y^j \to 0$ with any scaling factors $d_j \to 0$ is unique, i.e., $u_{y^j,d_j} \to u_0$.

Applying now Theorem 6.17 (that we prove later), we arrive at the following result.

Theorem 4.20. *Let $u \in P_1(M)$ be a solution of Problem* **A** *or* **B** *with 0 a regular point. Then there exists $\rho = \rho(u) > 0$ such that $\Gamma(u) \cap B_\rho$ is real analytic.*

4.4. Problem C: the free boundary near the branch points

In this section we study the structure of the free boundary near two-phase free boundary points in Problem **C**. By the implicit function theorem, we may restrict ourselves to the case of two-phase points where the gradient vanishes. Recall that these points are branch points (cf. Definition 3.24).

Theorem 4.21. *Let $u \in P_1(M)$ be a solution of Problem* **C** *such that* 0 *is a branch point, i.e.* $0 \in \Gamma_+(u) \cap \Gamma_-(u) \cap \Gamma^0(u)$. *Then there exists* $\rho = \rho(u) > 0$ *such that* $\Gamma_+(u) \cap B_\rho$ *and* $\Gamma_-(u) \cap B_\rho$ *are two touching* C^1 *graphs.*

The proof is based on the general idea of directional monotonicity that we exploited earlier.

Lemma 4.22. *Let $u \in P_1(M)$ be a solution of Problem* **C** *and suppose that for some* $C > 0$

$$C\partial_e u - |u| \geq -\varepsilon_0 \quad \text{in } B_1,$$

where $\varepsilon_0 < (1/8n)\min\{\lambda_+, \lambda_-\}$. *Then*

$$C\partial_e u - |u| \geq 0 \quad \text{in } B_{1/2}.$$

The proof is similar to those of Lemmas 4.3 and 4.13, but is subtler.

Proof. Let $v = C\partial_e u - |u|$ and suppose that the set $\{v < 0\}$ is nonempty. Note that $\nabla u \neq 0$ on $\gamma = \{v < 0\} \cap \{u = 0\}$ and therefore this set is locally a $C^{1,\alpha}$ surface. Further, note that

$$\Delta|u| \geq \lambda_+ \chi_{\{u>0\}} + \lambda_- \chi_{\{u<0\}} \quad \text{in } B_1,$$

in the sense of distributions, and

$$\Delta(\partial_e u) = (\lambda_+ + \lambda_-)(e \cdot \nu) H^{n-1} \lfloor \gamma \quad \text{in } \{v < 0\},$$

where $\nu = \nabla u / |\nabla u|$ is the unit normal on γ; see Exercise 3.6. Since on γ we have $u = 0$ and $v = C\partial_e u - |u| < 0$, we also have that $\partial_e u < 0$ and therefore $e \cdot \nu < 0$. In particular, this implies that

$$\Delta(\partial_e u) \leq 0 \quad \text{in } \{v < 0\},$$

and consequently

$$\Delta v \leq -\lambda_{\min} \quad \text{in } \{v < 0\},$$

where $\lambda_{\min} := \min\{\lambda_+, \lambda_-\}$. Suppose now that $\{v < 0\} \cap B_{1/2}$ contains a point x^0. Consider the auxiliary function

$$w(x) = v(x) + \frac{\lambda_{\min}}{2n}|x - x^0|^2.$$

Then $\Delta w \leq 0$ in $\{v < 0\}$ and $w(x^0) = v(x^0) < 0$. Hence, by the minimum principle

$$\inf_{\partial(B_{1/2}(x^0) \cap \{v<0\})} w \leq w(x^0) < 0.$$

Since $w(x) = \frac{\lambda_{\min}}{2n}|x - x^0|^2 \geq 0$ on $\partial\{v < 0\} \cap B_{1/2}(x^0)$, we must have

$$\inf_{(\partial B_{1/2}(x^0)) \cap \{v<0\}} w < 0.$$

Finally, since $w(x) = v(x) + \lambda_{\min}/8n$ on $(\partial B_{1/2}(x^0)) \cap \{v < 0\}$, we obtain that

$$\inf_{\partial B_{1/2}(x^0)} v < -\frac{\lambda_{\min}}{8n},$$

contrary to our assumption. □

Lemma 4.23. *Let $u \in P_1(M)$ with 0 a branch point and suppose that $u_0(x) = \frac{\lambda_+}{2}(x_n^+)^2 - \frac{\lambda_-}{2}(x_n^-)^2$ is a blowup of u at 0. Then for any $\delta \in (0, 1]$ there exists $r_\delta > 0$ such that*

$$(2\delta^{-1} r_\delta)\partial_e u - |u| \geq 0 \quad \text{in } B_{r_\delta},$$

for any unit vector $e \in \mathcal{C}_\delta = \{x : x_n > \delta|x'|\}$. In particular, $\partial_e u \geq 0$ in B_{r_δ}.

Proof. We start by observing that for u_0 we have

$$\partial_e u_0 = (e \cdot e_n)(\lambda_+ x_n^+ + \lambda_- x_n^-),$$

and thus

$$\delta^{-1}\partial_e u_0 - |u_0| \geq 0 \quad \text{in } B_1.$$

Consider now a sequence $r_j \to 0$ such that $u_{r_j} \to u_0$ in $C^{1,\alpha}_{\text{loc}}(\mathbb{R}^n)$. Then

$$\delta^{-1}\partial_e u_{r_j} - |u_{r_j}| \geq -\varepsilon_0 > -\frac{\lambda_{\min}}{8n} \quad \text{in } B_1,$$

for large j. But then by the previous lemma

$$\delta^{-1}\partial_e u_{r_j} - |u_{r_j}| \geq 0 \quad \text{in } B_{1/2}.$$

Letting $r_\delta = r_j/2$ and scaling back we obtain

$$(2\delta^{-1} r_\delta)\partial_e u - |u| \geq 0 \quad \text{in } B_{r_\delta}.$$

□

We are now ready to prove Theorem 4.21.

Proof of Theorem 4.21. 1) For $\delta \in (0, 1]$ let $r_\delta > 0$ be as in Lemma 4.23, so that $\partial_e u \geq 0$ in B_{r_δ} for any $e \in \mathcal{C}_\delta$. Then arguments, similar to Lemma 4.7 and Proposition 4.8, imply that $\Gamma_\pm(u) \cap B_{r_1/2}$ are graphs $x_n = f_\pm(x')$ of Lipschitz functions $f_\pm$ on $B'_{r_1/2}$ with $|\nabla_{x'} f_\pm| \leq \delta$ a.e. on $B'_{r_\delta/2}$ (see Exercises 4.1 and 4.4).

2) Now, to show the differentiability of $f_\pm$, observe that we have

$$|f_\pm(x') - f_\pm(0)| \leq \delta|x'|, \quad \text{for } |x'| \leq r_\delta/2.$$

By letting $\delta \to 0$ we easily obtain that $f_\pm$ are differentiable at 0 and $\nabla f_\pm(0) = 0$. Similarly, we obtain that $f_\pm$ are differentiable at any other branch point in $B_{r_1/2}$. The rest of the free boundary points $\hat{x} \in \Gamma \cap B_{r_1/2}$ are either (i) two-phase with $|\nabla u(\hat{x})| > 0$ or (ii) one-phase points. While the differentiability in case (i) is clear, in case (ii) we claim that $\hat{x}$ is a regular point. Indeed, if $\varepsilon > 0$ is such that $u \geq 0$ in $B_\varepsilon(\hat{x})$, then

$$\hat{x} - (\mathcal{C}_\delta \cap B_\varepsilon) \subset \{u = 0\}$$

and

$$\hat{x} + (\mathcal{C}_\delta \cap B_\varepsilon) \subset \{u > 0\},$$

similarly to Lemma 4.7. Hence, for a blowup $\hat{u}$ at $\hat{x}$ we must have

$$-\mathcal{C}_\delta \subset \{\hat{u} = 0\},$$

which is possible only if $\hat{u}$ is a half-space solution. Then we apply Theorem 4.10 to establish the differentiability of f_+ at $\hat{x}$.

Similarly, we treat the case when $u \leq 0$ near $\hat{x}$.

3) Finally, let us show that $f_\pm$ are C^1 functions. We first note that the inequality (see step 1) above)

$$|\nabla f_\pm(x')| \leq \delta, \quad \text{for } |x'| \leq r_\delta/2,$$

immediately implies

$$\lim_{x' \to 0} |\nabla f_\pm(x')| = 0 = |\nabla f(0)|.$$

Thus, the C^1 regularity at the origin follows.

Next, the C^1 regularity at other branch points in $B_{r_1/2}$ follows similarly. The C^1 regularity of $f_\pm$ at the remaining free boundary points has been actually established when we proved the differentiability at those points. This completes the proof of the theorem. □

4.5. Problem C: real analyticity of Γ^*

In general, C^1 is the best regularity one can get for the free boundaries $\Gamma_\pm$ near branch points in Problem **C**. We show this later in §6.3.2. Away from the branch points, Γ_+ and Γ_- either are separate, and then we treat them locally as free boundaries in the classical obstacle problem, or $|\nabla u|$ becomes nonzero, i.e. $\Gamma_\pm$ become Γ^*. The purpose of this section is to show that Γ^* is in fact locally a graph of a real analytic function.

Theorem 4.24. *Let u be a solution of Problem* **C** *in B^1 with $0 \in \Gamma^*(u)$. Then there exists $\rho = \rho(u) > 0$ such that $\Gamma^*(u) \cap B_\rho$ is real-analytic.*

Proof. The proof is performed by using the zeroth order partial hodograph-Legendre transformation; see e.g. [**KS80**, §VI.4]. We know that $u \in C^{1,\alpha}(B_1)$, $0 \in \Gamma^*(u)$, and therefore we can choose the coordinate system so that

$$\partial_{x_i} u(0) = 0, \quad i = 1, \dots, n-1, \quad \partial_{x_n} u(0) > 0.$$

Consider then the transformation of variables $y = T(x)$ given by

$$y_i = x_i, \quad i = 1, \dots, n-1, \quad y_n = u(x).$$

The Jacobian of this transformation is $\partial_{x_n} u$, which is nonzero at the origin, and hence T is invertible in a small ball B_ρ. Let

$$x_i = y_i, \quad i = 1, \dots, n-1, \quad x_n = v(y),$$

and define the inverse mapping $y = T^{-1}(x)$ on $T(B_\rho)$. Note that T takes $\Omega_\pm \cap B_\rho$ to $T(B_\rho) \cap U_\pm$, where $U_\pm = \{\pm y_n > 0\}$, and it takes $\Gamma \cap B_\rho$ to $T(B_\rho) \cap \{y_n = 0\}$. The smoothness of u in $\Omega_\pm$ implies the smoothness of v in $T(B_\rho) \cap U_\pm$. Moreover, we also have that $v \in C^{1,\alpha}(T(B_\rho))$. To obtain the equation satisfied by v we differentiate the identity $y_n = u(y', v(y))$ twice. This gives

$$\begin{aligned} -\Delta u &= (1 + |\nabla_{y'} v|^2) \frac{\partial_{y_n y_n} v}{(\partial_{y_n} v)^3} + \sum_{i=1}^{n-1} \frac{\partial_{y_i y_i} v}{\partial_{y_n} v} - 2 \sum_{i=1}^{n-1} \partial_{y_i y_n} v \frac{\partial_{y_i} v}{(\partial_{y_n} v)^2} \\ &= L(D_y^2 v, D_y v), \end{aligned}$$

which is a quasilinear uniformly elliptic operator in a neighborhood of the origin (see Exercise 4.6). We then obtain

$$Lv = L(D_y^2 v, D_y v) = \mp \lambda_\pm \quad \text{in } U_\pm \cap T(B_\rho).$$

Next, calling the part of function v in U_+ by v^1 and transforming the part of function v in U_- to a function v^2 in U_+ by an even reflection, we obtain a system

$$\begin{aligned} &Lv^1 = -\lambda_+, \quad Lv^2 = -\lambda_- \quad \text{in } U_+ \cap T(B_\rho), \\ &v^1 = v^2, \quad \partial_{y_n} v^1 = -\partial_{y_n} v^2 \quad \text{on } \partial U_+ \cap T(B_\rho). \end{aligned} \tag{4.7}$$

This system is a quasilinear elliptic system with coercive boundary conditions (see Exercise 4.7), and therefore we can apply [**Mor08**, Theorem 6.8.2] to conclude that v is real-analytic in $T(B_\rho) \cap \overline{U_+}$. Thus, we obtain that the free boundary Γ is given in B_ρ as the graph $x_n = v(x', 0)$ and therefore is real-analytic. □

Notes

The Lipschitz and C^1 regularity of the free boundary in the classical obstacle problem was proved for the first time by Caffarelli [**Caf77**]. The method of directional monotonicity used in this chapter is closely related to that of Alt [**Alt77**] for the Lipschitz regularity of the free boundary in a filtration problem (see also Friedman [**Fri88**, Chapter 2, Theorem 6.1]). Similar ideas, applied to pure second derivatives, can already be seen in Caffarelli [**Caf77**]. The observation that the directional monotonicity can be "inherited" from blowups to local solutions (as in Lemma 4.5) appeared in Caffarelli [**Caf98a**].

For Problem **A**, the directional monotonicity was proved by Caffarelli-Karp-Shahgholian [**CKS00**]. For Problem **B**, a similar property of solutions was proved by Caffarelli-Salazar-Shahgholian [**CSS04**], and for Problem **C**, it is due to Shahgholian-Uraltseva-Weiss [**SUW07**]. Our exposition mainly follows the latter four papers.

In the classical obstacle problem, there is a sharp result due to Blank [**Bla01**] that the minimal regularity of the obstacle to guarantee the C^1 regularity of the free boundary (near regular points) is the Dini continuity of its Laplacian. This result was extended to the no-sign obstacle problem (see §1.1.2) with double-Dini continuous f by Petrosyan-Shahgholian [**PS07**], and recently to Dini continuous f by Andersson-Lindgren-Shahgholian [**ALS12**], matching the regularity in the classical case. This was achieved by establishing the $C^{1,1}$ regularity of the solutions to that problem.

For remarks regarding the partial hodograph-Legendre transformation, we refer to the Notes at the end of Chapter 6.

The parabolic counterparts of many results in this chapter can be found in Caffarelli-Petrosyan-Shahgholian [**CPS04**] for Problem **A** and in Shahgholian-Uraltseva-Weiss [**SUW09**] for Problem **C**.

Exercises

4.1. Let $u \in C^1(B_1)$ be such that

$$\partial_e u \geq 0 \quad \text{in } B_1, \quad \text{for any } e \in \mathcal{C}_\delta,$$

for some $\delta > 0$. Suppose that $\{u = 0\} \neq \emptyset$. Prove that there are two functions $f_\pm$ on B_1' satisfying the Lipschitz condition with the constant not exceeding δ such that $f_+ \geq f_-$ and

$$\{u > 0\} = \{x \in B_1 : x_n > f_+(x')\},$$
$$\{u < 0\} = \{x \in B_1 : x_n < f_-(x')\}.$$

Hint: For fixed $\rho < 1$ consider two points x and y contained in $\partial\{u = 0\} \cap B_\rho$. Check that if $y_n - x_n > \delta|y' - x'|$, then $y \in \partial\{u > 0\}$, $x \in \partial\{u < 0\}$, $y \notin \partial\{u < 0\}$, $x \notin \partial\{u > 0\}$.

4.2. Complete the proof of Proposition 4.8.

Hint: To show that $|\nabla_{x'} f| \leq \delta$ a.e. on $B'_{r_\delta/2}$ note that $(x', f(x')) \in B_{r_\delta}$ whenever $x' \in B'_{r_\delta/2}$.

4.3. Complete the proof of Lemma 4.13.

4.4. State and prove the analogue of Proposition 4.8 for solutions of Problem **C** with a branch point at the origin.

4.5. Let Ω be a certain open set in $\mathbb{R}^n$ and h a continuous function on $\overline{\Omega} \cap B_1$ such that

1) $h \geq 0$ on $\partial\Omega \cap B_1$;
2) $\Delta h \leq -1$ in $\Omega \cap B_1$;
3) $h \geq -\varepsilon$ in $\Omega \cap B_1$.

Prove that if $\varepsilon < 1/8n$, then $h \geq 0$ in $\Omega \cap B_{1/2}$.

Verify that Lemmas 4.3, 4.13, and 4.22 (with some work) are particular cases of this problem.

4.6. Derive the equation for $L(D_y^2 v, D_y v) = 0$ in the proof of Theorem 4.24 and show that it is uniformly elliptic in a neighborhood of the origin.

Hint: Differentiate the identity $y_n = u(y', v(y))$ to obtain

$$1 = \partial_{x_n} u \, \partial_{y_n} v, \quad 0 = \partial_{x_i} u + \partial_{x_n} u \, \partial_{y_i} v, \quad i = 1, \ldots, n-1$$

in $T(B_\rho)$. Differentiating one more time in $T(B_\rho) \cap U_\pm$, we then have

$$0 = \partial_{x_n x_n} u (\partial_{y_n} v)^2 + \partial_{x_n} u \, \partial_{y_n y_n} v,$$
$$0 = \partial_{x_n x_i} u \, \partial_{y_n} v + \partial_{x_n x_n} u \, \partial_{y_i} v \, \partial_{y_n} v + \partial_{x_n} \partial_{y_n y_i} v,$$
$$0 = \partial_{x_i x_i} u + 2\partial_{x_n x_i} u \partial_{y_i} v + \partial_{x_n x_n} u (\partial_{y_i} v)^2 + \partial_{x_n} u \, \partial_{y_i y_i} v.$$

4.7. Show that (4.7) is a quasilinear elliptic system with coercive boundary conditions with the choices $s_1 = s_2 = 0$, $t_1 = t_2 = 2$, $r_1 = -2$, $r_2 = -1$ (see [**KS80**, §VI.3] for definitions).

Chapter 5

Global solutions

In this chapter we study the so-called global solutions, i.e. solutions defined in the entire space $\mathbb{R}^n$, with an additional assumption that they grow quadratically at infinity. More precisely, we consider elements of the class $P_\infty(x^0, M)$ which satisfy

- $\|D^2 u\|_{L^\infty(\mathbb{R}^n)} \le M$,
- $x^0 \in \Gamma(u)$ in Problems **A**, **B**; $x^0 \in \Gamma^0(u)$ in Problem **C**.

The main motivation for the study of global solutions is that they appear as blowups of a single function or a sequence of functions with variable centers, i.e. the limits of rescalings

$$u^j_{x^j, r_j}(x) = \frac{u^j(x^j + r_j x) - u^j(x^j)}{r_j^2}$$

as $r_j \to 0$. Recall that when the functions u^j and the centers x^j are fixed, we can give a complete characterization of such blowups in all of our model problems; see Theorem 3.22. However, Theorem 3.22 is no longer applicable if either u^j or x^j are variable, and essentially the only information we obtain is that the limit is from the class $P_\infty(M)$.

We will first study the global solution for the classical obstacle problem in §5.1, then generalize the results for Problems **A**, **B** in §5.2. In §5.3 we consider Problem **C** and complete the chapter by discussing approximations of solutions defined in large balls by global solutions (§5.4).

5.1. Classical obstacle problem

Theorem 5.1. *Let $u \in P_\infty(M)$ be a global solution of the classical obstacle problem $\Delta u = \chi_{\{u>0\}}$, $u \geq 0$, in $\mathbb{R}^n$. Then u is a convex function in $\mathbb{R}^n$, i.e.*

$$\partial_{ee} u(x) \geq 0, \quad \textit{for any direction } e \textit{ and } x \in \mathbb{R}^n.$$

In particular, the coincidence set $\{u = 0\}$ is convex.

Proof. Fix any direction e. Since we know that $\partial_{ee} u$ is an L^∞ function vanishing a.e. on $\{u = 0\}$, it will be enough to show that $\partial_{ee} u \geq 0$ in $\Omega(u) = \{u > 0\}$. Without loss of generality, assume $e = e_n = (0, \dots, 0, 1)$. Then, arguing by contradiction, suppose that

$$-m := \inf_{\Omega(u)} \partial_{x_n x_n} u < 0,$$

and let $x^j \in \Omega(u)$ be a minimizing sequence for the value $-m$, i.e.

$$\lim_{j\to\infty} \partial_{x_n x_n} u(x^j) = -m.$$

Let $d_j = \operatorname{dist}(x^j, \Gamma(u))$ and consider the rescalings

$$u_j(x) = u_{x^j, d_j}(x) = \frac{1}{d_j^2} u(x^j + d_j x).$$

Observe that $B_1 \subset \Omega(u_j)$ and the free boundary $\Gamma(u_j)$ contains at least one point on ∂B_1. Since also $|D^2 u_j|$ are uniformly bounded by M, we have the uniform estimates

$$|u_j(x)| \leq \frac{M}{2}(|x| + 1)^2,$$

and therefore we can extract a subsequence converging in $C^{1,\alpha}_{\mathrm{loc}}(\mathbb{R}^n)$ to a global solution $u_0 \geq 0$ of the classical obstacle problem. Moreover, from the respective properties of u_j, we obtain that $\Delta u_0 = 1$ in B_1 and that ∂B_1 contains at least one free boundary point of u_0. Carefully observe here that we cannot claim that $B_1 \subset \{u_0 > 0\}$, but we readily have that the set $\{u = 0\} \cap B_1$ has zero measure. For the rest of the proof, we will denote $\tilde{\Omega}(u_0) = B_1 \cup \{u_0 > 0\}$. It is easy to see that $\Delta u_0 = 1$ in $\tilde{\Omega}(u_0)$ in the classical sense.

Since all functions u_j satisfy $\Delta u_j = 1$ in B_1, the convergence to u_0 can be assumed to be at least in $C^2_{\mathrm{loc}}(B_1)$. Hence, the limit function u_0 satisfies

$$\Delta u_0 = 1, \qquad \partial_{x_n x_n} u_0 \geq -m \quad \text{in } B_1, \qquad \partial_{x_n x_n} u_0(0) = -m.$$

Since $\partial_{x_n x_n} u_0$ is harmonic in B_1, the minimum principle implies that

$$\partial_{x_n x_n} u_0 \equiv -m \quad \text{in } B_1.$$

In fact, we have $\partial_{x_n x_n} u_0 = -m$ in the connected component of $\tilde{\Omega}(u_0)$ which contains B_1. Hence we obtain the representations

$$\partial_{x_n} u_0(x) = g_1(x') - m x_n, \quad x' = (x_1, \dots, x_{n-1}), \tag{5.1}$$

and

$$u_0(x) = g_2(x') + g_1(x') x_n - \frac{m}{2} x_n^2, \tag{5.2}$$

in B_1. Now, choose a point $(x', 0) \in B_1$ and start moving in the direction e_n. Observe that as long as we stay in $\tilde{\Omega}(u_0)$, we still have $\partial_{x_n x_n} u = -m$ and therefore still have the representations (5.1)–(5.2). However, sooner or later we will reach $\partial\tilde{\Omega}(u_0)$; otherwise, if x_n becomes too large, (5.2) will imply $u_0 < 0$, contrary to our assumption. Since $u_0 = |\nabla u_0| = 0$ on $\partial\tilde{\Omega}(u_0)$, from (5.1) we obtain that the first value $\xi(x')$ of x_n for which we arrive at $\partial\tilde{\Omega}(u_0)$ is given by

$$\xi(x') = \frac{g_1(x')}{m}.$$

Hence from (5.2) we deduce that

$$g_2(x') = -\frac{g_1(x')^2}{2m}.$$

Now, the representation (5.2) takes the form

$$u_0(x) = -\frac{m}{2}(x_n - \xi(x'))^2,$$

which is not possible since $u_0 \geq 0$. This concludes the proof. □

An alternative proof of Theorem 5.1 is given in Exercise 5.1, by using the method of [**Caf98a**], which goes back to [**Caf77**].

5.2. Problems A, B

In this section we give a characterization of global solutions of Problems **A** and **B**.

Theorem 5.2. *Let $u \in P_\infty(M)$ be a global solution of Problem* **A** *or* **B**. *Then either u is a quadratic polynomial, or u is convex. In the latter case there exists a constant a such that $u \geq a$ and $\Omega^c(u) = \{u = a\}$.*

Note that the constant a must be zero in the case of Problem **A** and $a = u(0)$ in Problem **B**; see Exercise 5.6.

The proof of the theorem will be split into two parts, depending on whether $\Omega^c(u)$ is bounded or not. But first, we establish certain properties of *shrinkdowns*. Recall that for a global solution $u \in P_\infty(x^0, M)$ of Problem **A**, **B**, or **C**, shrinkdowns are the limits

$$u_\infty(x) = \lim_{R_j\to\infty} u_{x^0,R_j}(x) \quad \text{in } W^{2,p}_{\mathrm{loc}}(\mathbb{R}^n),\ 1 < p < \infty,$$

where

$$u_{x^0,R_j}(x) = \frac{u(x^0 + R_j x) - u(x^0)}{R_j^2}$$

and $R_j \to \infty$ is an arbitrary sequence. Note that contrary to blowups, shrinkdowns are defined only for global solutions. Moreover, it is easy to see that the shrinkdowns are independent of the choice of x^0 on Γ (or even in $\mathbb{R}^n$) and there is even no need in subtracting $u(x^0)$; shrinkdowns depend only on the sequence $R_j \to \infty$ (see Exercise 5.7).

Similar to the classification of blowups in Theorem 3.22, we have an analogous classification of shrinkdowns.

Proposition 5.3 (Shrinkdowns: Problems **A**, **B**). *Let $u \in P_\infty(x^0, M)$ be a global solution of Problem* **A** *or* **B**. *Then any shrinkdown u_∞ of u at x^0 is ether a half-space or a polynomial solution (as defined in Theorem* 3.22*).*

See also Proposition 5.9 for a similar statement for Problem **C**, with an identical proof.

Proof. The proof is almost the same as that of Theorem 3.18. In this case, however, we notice that the function

$$r \mapsto \phi_e(r, u, x^0) = \Phi(r, (\partial_e u)^\pm, x_0)$$

is monotone, so that the limit $\phi_e(+\infty, u, x^0)$ exists. The reason is that we are restricted here to Problems **A** and **B**, for which $(\partial_e u)^\pm$ are subharmonic (see Remark 2.16), and we can use Theorem 2.4 instead of Theorem 2.13. Then we have

$$\phi_e(r, u_\infty, 0) \equiv \phi_e(+\infty, u, x^0)$$

and we can proceed as in Theorem 3.18. □

Proposition 5.4. *Let u and x^0 be as in Proposition* 5.3. *If there exists a shrinkdown u_∞ of u at x^0 which is a monotone function of one variable, then u itself is a monotone function of one variable.*

Proof. If u_∞ is a monotone function of one variable, then for any direction $e \in \mathbb{R}^n$, one of the functions $(\partial_e u_\infty)^+$ and $(\partial_e u_\infty)^-$ vanishes identically in $\mathbb{R}^n$ and therefore $\phi_e(r, u_\infty, 0) \equiv 0$. But then

$$0 \le \phi_e(r, u, x^0) \le \phi_e(+\infty, u, x^0) = \phi_e(1, u_\infty, 0) = 0.$$

Hence, $\phi_e(r, u, x^0) = 0$ for any $r > 0$, which implies the statement of the proposition. □

Remark 5.5. We explicitly observe here that for solutions of Problem **A**, or solutions of Problem **B** with $u(x^0) = \sup_{\Omega^c(u)} u$, the results above can be proved with the help of Weiss's monotonicity formula. In fact, in that case one can prove also a statement which is in a sense dual to the one above: If a blowup of a global solution u at x^0 is polynomial, then u itself is polynomial. See the proof of Lemma 5.6 below.

5.2.1. The compact complement case.

For the proof of Theorem 5.2, we first consider the case when $\Omega^c(u)$ is compact.

Lemma 5.6. *Let $u \in P_\infty(x^0, M)$ be a global solution of Problem* **A** *or that of Problem* **B** *with an additional assumption $u(x^0) = \sup_{\Omega^c(u)} u$. If $\Omega^c(u)$ is compact and $\operatorname{Int}\Omega^c(u) \neq \emptyset$, then x^0 is a regular point.*

Proof. First, note that because of our assumptions, Weiss's functional $r \mapsto W(r, u, x^0)$ is nondecreasing; see Theorem 3.26. Next, arguing by contradiction, suppose that x^0 is a singular point. Further, let u_0 be a blowup of u at x^0 and u_∞ a shrinkdown. Then we have

$$W(1, u_0) = W(0+, u, x^0) \leq W(r, u, x^0) \leq W(+\infty, u, x^0) = W(1, u_\infty).$$

On the other hand, since u_0 is polynomial, we have $W(1, u_0) = \alpha_n$, while $W(1, u_\infty)$ in general may take only two values: $\alpha_n/2$ or α_n; see Lemma 3.31 (this lemma is applicable, since Problems **A** and **B** have the same homogeneous global solutions). Thus, the only possibility is to have

$$W(r, u, x^0) \equiv \alpha_n.$$

This implies that $u - u(x_0)$ is homogeneous of degree two and consequently $u(x) = u(x_0) + u_0(x)$, which implies that u is a quadratic polynomial. The latter is impossible, since $\operatorname{Int}\Omega^c(u) \neq \emptyset$. □

Lemma 5.7. *Let u be as in Lemma 5.6 and suppose that*

$$\sup_{\Omega^c} u = 0. \tag{5.3}$$

Then there exists $\zeta^0 \in \Omega^c$ such that the function

$$r \mapsto \frac{u(\zeta^0 + rx)}{r^2}$$

is nondecreasing, for any fixed x.

Proof. We will give the proof for $n \geq 3$. Denote by V the Newtonian potential of Ω^c, i.e.

$$V(x) = \int_{\Omega^c} \frac{c_n}{|x - y|^{n-2}} dy,$$

where $c_n > 0$ is chosen so that $\Delta V = -\chi_{\Omega^c}$ in the sense of distributions. Then V is bounded and superharmonic in $\mathbb{R}^n$ and harmonic in Ω. By the maximum principle, there is at least one point $\zeta^0 \in \Omega^c$ such that

$$V(\zeta^0) \geq V(x), \quad \text{for all } x \in \mathbb{R}^n.$$

Without loss of generality, we may assume that $\zeta^0 = 0$. Since

$$\Delta(u - V) = 1 \quad \text{in } \mathbb{R}^n,$$

in the sense of distributions, and all second order partial derivatives of $u - V$ are bounded harmonic functions, the Hessian of $u - V$ is a constant matrix, by the Liouville theorem. Hence $u - V$ is a polynomial of degree two. Set

$$P(x) = u(x) - V(x) - u(0) + V(0).$$

Note that by construction $|\nabla V(0)| = |\nabla u(0)| = 0$. Hence $P(0) = |\nabla P(0)| = 0$, which implies that P is homogeneous of degree two. Now consider the function

$$h(x) = x \cdot \nabla u(x) - 2u(x).$$

Then h is continuous in $\mathbb{R}^n$ and for all $x \neq 0$ fixed,

$$\frac{d}{dr}\left(\frac{u(rx)}{r^2}\right) = \frac{1}{r^3} h(rx).$$

Thus, to complete the proof of the lemma, it will suffice to show that $h \geq 0$ in $\mathbb{R}^n$. First notice that

$$h(x) = -2u(x) \geq 0, \quad \text{for all } x \in \Omega^c,$$

so it remains to show that $h \geq 0$ in Ω. To this end, note that by a simple calculation h is harmonic in Ω and therefore we may invoke the maximum principle and conclude that $h \geq 0$ in Ω, provided we know that

$$h \geq 0 \quad \text{on } \partial\Omega,$$

and (since Ω is a complement of a bounded set)

$$\liminf_{|x| \to \infty} h(x) \geq 0.$$

While we already know the former inequality, since $h \geq 0$ on Ω^c, we need to prove the latter. From the definition of h and using that $u = P + V + u(0) - V(0)$ we have

$$h(x) = [x \cdot \nabla P(x) - 2P(x)] + [x \cdot \nabla V(x) - 2V(x)] + 2V(0) - 2u(0).$$

The first term in brackets is identically zero, since P is homogeneous of degree 2. Further, from compactness of Ω^c, it is easy to see that

$$\lim_{|x| \to \infty} [x \cdot \nabla V(x) - 2V(x)] = 0.$$

Finally, $V(0) \geq 0$ and $u(0) \leq 0$ by construction. Therefore, we obtain

$$\lim_{|x|\to\infty} h(x) = 2V(0) - 2u(0) \geq 0.$$

Thus, we obtain that $h \geq 0$ in Ω and consequently in $\mathbb{R}^n$, completing the proof of the lemma. □

Now we can prove Theorem 5.2 in the case of the compact complement.

Proof of Theorem 5.2: Ω^c compact. Without loss of generality we may assume that $\operatorname{Int}\Omega^c \neq \emptyset$ (or equivalently $|\Omega^c| > 0$); otherwise u is a polynomial solution and we are done.

To proceed, we consider Problems **A** and **B** separately. In both cases, we set the origin at ζ^0 found in Lemma 5.7.

Problem **A**. We claim that there exists a small $\rho > 0$ such that $u \geq 0$ in B_ρ. Indeed, if $0 \in \operatorname{Int}\Omega^c$, this is immediate. If $0 \in \Gamma$, then it is regular by Lemma 5.6 and therefore the claim follows from Lemma 4.5. Now, invoking Lemma 5.7, we conclude that

$$0 \leq u(\rho x) \leq \frac{\rho^2}{R^2} u(Rx), \quad x \in B_1, \ R > \rho,$$

i.e. $u \geq 0$ everywhere in $\mathbb{R}^n$. After that we apply Theorem 5.1.

Problem **B**. Without loss of generality we may assume that (5.3) is satisfied. Then the set $U := \{u \leq 0\}$ is starshaped with respect to the origin, and we also have that $\Omega^c \subset U$. Since Ω^c is compact and u is locally constant in the interior of Ω^c, there exists $x^0 \in \partial\Omega^c = \Gamma$ such that $u(x^0) = 0$. By Lemma 5.6, x^0 is a regular point. In particular, by Lemma 4.17 and Theorem 4.11, u vanishes on a truncated cone $x^0 - (\mathcal{C}_{2\delta} \cap B_{r'_\delta}) \subset \Omega^c \subset U$ (after a possible rotation of coordinate axes). Now take a small ball $B \subset x^0 - (\mathcal{C}_{2\delta} \cap B_{r'_\delta})$ and consider the convex hull $\mathcal{K}$ of B and $\{0\}$. Note that $\mathcal{K} \subset U$ from the starshapedness of U. Moreover, since $u = 0$ in B, $u \leq 0$ and subharmonic in $\mathcal{K}$, and since $\mathcal{K}$ is connected, u must vanish identically in $\mathcal{K}$. This also implies that $\mathcal{K} \subset \Omega^c$. As a consequence, we obtain that

$$u(0) = 0.$$

We now claim that $0 \in \operatorname{Int}\{u \geq 0\}$. Once we show that, we will immediately obtain that $u \geq 0$ in $\mathbb{R}^n$ and the problem will be reduced to Theorem 5.1. Consider two possibilities:

1) $0 \in \operatorname{Int}\Omega^c$. Then $u \equiv u(0) = 0$ in a neighborhood of the origin, so the claim follows trivially.

2) $0 \in \partial\Omega^c$. In this case Lemma 5.6 implies that 0 is a regular point. Then Theorem 4.11 implies that $u \equiv 0$ on $\Omega^c \cap B_\rho$ for some small $\rho > 0$.

This means that locally it is Problem **A**, and by Lemma 4.5 we have that $u \geq 0$ in $B_{\rho'}$ for some $\rho' > 0$.

This proves the claim and completes the proof of the theorem. $\square$

5.2.2. Global solutions with unbounded Ω^c. Here we complete the proof of Theorem 5.2 by considering the remaining case of unbounded Ω^c. The proof uses a dimension reduction argument.

Proof of Theorem 5.2: Ω^c unbounded. Again, as before, we may assume that $\operatorname{Int}\Omega^c \neq \emptyset$. We proceed under this assumption.

First, suppose that a shrinkdown u_∞ of u at 0 is a half-space solution. Then, by Proposition 5.4 we have that $u - u(0)$ is a half-space solution and the theorem follows in this case.

Next, consider the case when no shrinkdown of u at 0 is a half-space solution. Then all shrinkdowns are polynomial by Proposition 5.3. Note that by our assumption (that $\operatorname{Int}\Omega^c \neq \emptyset$) u itself cannot be polynomial.

By our assumption Ω^c is unbounded. Then, it is easy to see that Ω must also be unbounded. Therefore, there exists a sequence $x^j \in \partial\Omega$ such that $R_j = |x^j| \to \infty$. Consider then the limit u_∞ of rescalings u_{0,R_j}, passing to a subsequence, if necessary. We may also assume that $e^j = x^j/R_j$ converge to a point e^0 on the unit sphere. Then $e^0 \in \partial\Omega(u_\infty)$. Moreover, by homogeneity, $\mathbb{R}^+e^0 = \{re^0 \colon r > 0\}$ must lie completely in the free boundary $\partial\Omega(u_\infty)$. Then again, since u_∞ is a homogeneous quadratic polynomial, this is possible only if $\partial_{e^0}u_\infty \equiv 0$ (verify this!). Consider now the Alt-Caffarelli-Friedman functional

$$\phi_{e^0}(r, u) = \Phi(r, (\partial_{e^0}u)^+, (\partial_{e^0}u)^-).$$

Recall that by Lemma 2.15 $(\partial_{e^0}u)^\pm$ are subharmonic (see Remark 2.16), and therefore we may apply the ACF monotonicity formula (Theorem 2.4) to obtain that

$$0 \leq \phi_{e^0}(r, u) \leq \phi_{e^0}(\infty, u) = \phi_{e^0}(1, u_\infty) = 0.$$

Hence, either $(\partial_{e^0}u)^+$ or $(\partial_{e^0}u)^-$ must vanish identically and we may assume that $\partial_{e^0}u \geq 0$ (otherwise we replace e^0 by $-e^0$).

Next, by rotation invariance, we may also assume that $e^0 = e_n$. Besides, since $\operatorname{Int}\Omega^c$ is nonempty, we may assume that

$$B_\delta \subset \Omega^c(u)$$

for some $\delta > 0$, after a possible change of the origin. Then $u \equiv u(0)$ in B_δ. Moreover, by the monotonicity in the direction e_n we have that

$u \le u(0)$ in the half-infinite cylinder $B'_\delta \times (-\infty, 0)$. Since u is subharmonic, the maximum principle implies now that

$$u \equiv u(0) \quad \text{in } B'_\delta \times (-\infty, 0].$$

Define an $(n-1)$-dimensional function

$$\hat{u}(x') = \lim_{x_n \to -\infty} u(x', x_n) - u(0).$$

First, we notice that the limit exists for any $x' \in \mathbb{R}^{n-1}$ by the monotonicity in the direction e_n. Next, the limit is finite, since $B'_\delta \times (-\infty, 0] \subset \Omega^c$, which gives the estimate

$$|u(x', x_n) - u(0)| \le \frac{M}{2}|x'|^2.$$

Thus, in the limit, we obtain that

$$|\hat{u}(x')| \le \frac{M}{2}|x'|^2, \quad x' \in \mathbb{R}^{n-1}.$$

Further, we claim that $\hat{u}(x')$ solves the same problem as u, but in $\mathbb{R}^{n-1}$. Indeed, if we add a dummy variable x_n, then $(x', x_n) \mapsto \hat{u}(x')$ is a limit of solutions $u(x', x_n - t)$ as $t \to \infty$ and therefore is a solution in $\mathbb{R}^n$. But then it is equivalent to saying that $\hat{u}$ is a solution in $\mathbb{R}^{n-1}$.

To conclude the proof, we make an inductive assumption that Theorem 5.2 holds in dimension $(n-1)$.

We thus may apply Theorem 5.2 to $\hat{u}$. Note that

$$\hat{u} = 0 \quad \text{in } B'_\delta$$

and therefore $\hat{u}$ cannot be polynomial and therefore must be convex and satisfy $\hat{u}(x') \ge \hat{u}(0) = 0$. But then

$$u(x', x_n) - u(0) \ge \hat{u}(x') \ge 0.$$

Thus, $u - u(0)$ is a global solution of the classical obstacle problem (see Exercise 5.6), and we conclude the proof by applying Theorem 5.1.

Finally, to complete the induction argument, we need to verify Theorem 5.2 in dimension one. This is easy to do: up to an additive constant the one-dimensional global solutions have one of the forms

$$\frac{1}{2}[(x_1 - a)^-]^2, \quad \frac{1}{2}[(x_1 - b)^+]^2, \quad \frac{1}{2}[(x_1 - a)^-]^2 + \frac{1}{2}[(x_1 - b)^+]^2,$$

with $a \le b$ in the latter case. Obviously, Theorem 5.2 is satisfied for these solutions, and this completes our proof. □

5.3. Problem C

In Problem **C**, just a mere fact that u has a branch point, makes u a two-half-space solution.

Theorem 5.8. *Let $u \in P_\infty(M)$ be a solution of Problem* **C** *such that the origin is a branch point, i.e. $0 \in \Gamma_+(u) \cap \Gamma_-(u) \cap \Gamma^0(u)$. Then u is a two-half-space solution*

$$u(x) = \frac{\lambda_+}{2}[(x \cdot e)^+]^2 - \frac{\lambda_-}{2}[(x \cdot e)^-]^2,$$

for a certain direction e.

To prove this theorem, we start with the general classification of shrinkdowns.

Proposition 5.9 (Shrinkdowns: Problems **C**). *Let $u \in P_\infty(x^0, M)$ be a global solution of Problem* **C**. *Then any shrinkdown u_∞ is ether a half-space solution, a positive or negative polynomial, or a two-half-space solution (see Theorem 3.22).*

Proof. See the proof of Proposition 5.3. □

Now we are ready to prove Theorem 5.8.

Proof of Theorem 5.8. Consider a shrinkdown u_∞ of u at the origin. Because of the nondegeneracy and $C^{1,\alpha}$ convergence of rescalings, it is easy to see that 0 is a two-phase boundary point for u_∞. On the other hand, by Proposition 5.9 above, u_∞ is either a half-space, a positive or negative polynomial, or a two-half-space solution. But only for two-half-space solution the origin is a two-phase boundary point. Therefore, u_∞ is a two-half-space solution for a certain direction e^0.

In particular, for any direction e, $\partial_e u_\infty$ does not change sign in $\mathbb{R}^n$ and therefore

$$\phi_e(r, u_\infty) = \Phi(r, (\partial_e u_\infty)^+, (\partial_e u_\infty)^-) = 0.$$

On the other hand, since $(\partial_e u)^\pm$ are subharmonic (see Remark 2.16), the function $r \mapsto \phi_e(r, u)$ is monotone by the ACF monotonicity formula (Theorem 2.4) and we have

$$0 \leq \phi_e(r, u) \leq \phi_e(+\infty, u) = \phi_e(1, u_\infty) = 0,$$

implying that $\partial_e u$ does not change sign. Since this holds for all directions e, we conclude (see Exercise 5.4) that u is one-dimensional and hence can be computed as before. □

5.4. Approximation by global solutions

Once we classified the global solutions, we want to show that the solutions in large domains can be approximated by global solutions and that during this process certain properties of global solutions are retained.

5.4.1. Problems A and B.

Lemma 5.10 (Approximation by global solutions). *For any $\varepsilon > 0$ there exists a radius $R_\varepsilon = R_\varepsilon(M, n)$ such that if $u \in P_R(M)$ with $R \geq R_\varepsilon$, then one can find a global solution $u_0 \in P_\infty(M)$ such that*

$$\|u - u_0\|_{C^1(B_1)} \leq \varepsilon.$$

Proof. Fix $\varepsilon > 0$ and argue by contradiction. If no such R_ε exists, then one can find a sequence $R_j \to \infty$ and solutions $u_j \in P_{R_j}(M)$ such that

$$\|u_j - u_0\|_{C^1(B_1)} > \varepsilon \quad \text{for any } u_0 \in P_\infty(M). \tag{5.4}$$

Note that in the case of Problem **B** we may additionally ask for $u_j(0) = 0$. Then we will have uniform estimates

$$|u_j(x)| \leq \frac{M}{2}|x|^2, \quad |x| \leq R_j.$$

We may assume that u_j converges is $C^1_{\mathrm{loc}}(\mathbb{R}^n)$ to a global solution $\hat{u}_0 \in P_\infty(M)$. But then

$$\|u_j - \hat{u}_0\|_{C^1(B_1)} < \varepsilon, \quad \text{for } j \geq j_\varepsilon,$$

contradicting (5.4). □

Next, we want to find conditions on local solutions to Problems **A**, **B** that will guarantee approximation by global solutions for which the origin is regular.

Definition 5.11 (Minimal diameter). The *minimal diameter* of a bounded set $E \subset \mathbb{R}^n$, denoted $\operatorname{min\,diam}(E)$, is the infimum of distances between pairs of parallel planes enclosing E.

Definition 5.12 (Thickness function). For $u \in P_R(x^0, M)$, the *thickness function* of Ω^c at x^0 is defined by

$$\delta(\rho, u, x^0) = \frac{\operatorname{min\,diam}(\Omega^c \cap B_\rho(x^0))}{\rho}, \quad 0 < \rho < R.$$

As usual, we drop x^0 from the notation in the case $x^0 = 0$.

Lemma 5.13 (Approximation by convex global solutions). *Fix $\sigma \in (0, 1]$ and $\varepsilon > 0$. Then there exists $R_{\varepsilon,\sigma} = R_{\varepsilon,\sigma}(M, n)$ such that if*

- *$u \in P_R(M)$ for $R \geq R_{\varepsilon,\sigma}$ and*
- *$\delta(1, u) \geq \sigma$,*

then we can find a convex global solution $u_0 \in P_\infty(M)$ *with the properties*

(i) $\|u - u_0\|_{C^1(B_1)} \leq \varepsilon$;

(ii) *there exists a ball* $B = B_\rho(x) \subset B_1$ *of radius* $\rho = \sigma/2n$ *such that* $B \subset \Omega^c(u_0)$.

Moreover, if u_0 *and* $B_\rho(x)$ *are as above and* $\varepsilon \leq \varepsilon(\sigma)$, *then*

(iii) $B_{\rho/2}(x) \subset \Omega^c(u)$.

Proof. As before, we argue by contradiction. If no $R_{\varepsilon,\sigma}$ exists as above, then we can find a sequence $R_j \to \infty$ and solutions $u_j \in P_{R_j}(M)$ such that

$$\delta(1, u_j) \geq \sigma,$$

and that for any global solution $u_0 \in P_\infty(M)$ at least one of conditions (i)–(ii) in the lemma fails.

Now, we again have uniform estimates

$$|u_j(x)| \leq \frac{M}{2}|x|^2, \quad |x| \leq R_j,$$

and therefore without loss of generality we may assume that u_j converges in $C^1_{\mathrm{loc}}(\mathbb{R}^n)$ to a global solution $\hat{u}_0 \in P_\infty(M)$. Thus,

$$\|u_j - \hat{u}_0\|_{C^1(B_1)} < \varepsilon \quad \text{for } j \geq j_\varepsilon,$$

i.e. condition (i) in the lemma is satisfied.

To show that condition (ii) is also satisfied with this choice of the global solution, we note that

$$\delta(1, \hat{u}_0) \geq \sigma.$$

This can be shown by contradiction: if $\delta(1, \hat{u}_0) < \sigma$, then $\delta(1, u_j) < \sigma$ for large j (see Exercise 5.5). This implies that $\hat{u}_0$ cannot be a polynomial solution. Consequently, by the classification of global solutions, $\hat{u}_0$ is a convex function and $\Omega^c(\hat{u}_0)$ is a convex set.

To proceed, we will need the following lemma of John [**Joh48**] on convex sets.

Lemma 5.14 (John's ellipsoid lemma). *If* C *is a convex body in* $\mathbb{R}^n$ *and* E *is the ellipsoid of largest volume contained in* C, *then* $C \subset nE$ *(after we choose the origin at the center of* E*).*

Applying this lemma to the convex set $C = \Omega^c(\hat{u}_0) \cap B_1$, we obtain that C contains a ball B of radius $\rho = \sigma/2n$. Indeed, if E has one of its diameters smaller than 2ρ, then nE has one of its diameters smaller than $2n\rho = \sigma$ and C is contained in a strip of width smaller than σ, a contradiction.

Thus, (ii) is also satisfied for $\hat{u}_0$, which means we have arrived at a contradiction with the assumption that no $R_{\varepsilon,\sigma}$ exists.

Finally, (iii) follows from the nondegeneracy (see Lemma 3.17(ii)). □

5.4.2. Problem C. In the case of the two-phase obstacle problem, we specify a condition that guarantees the approximation by two-half-space solutions. For our application, we will need to do that for a class of local solutions slightly more general than $P_R(M)$.

Lemma 5.15 (Approximation by two-half-space solutions). *For any $\varepsilon > 0$ there exist $R_\varepsilon = R_\varepsilon(M, n)$ and $\sigma_\varepsilon = \sigma_\varepsilon(M, n)$ such that if u is a solution of Problem **C** in B_R for $R \geq R_\varepsilon$, with $\|D^2 u\|_{L^\infty(B_R)} \leq M$, such that*

- $|\nabla u(0)| \leq \sigma_\varepsilon$ *and*
- $\operatorname{dist}(0, \{u > 0\}) \leq \sigma_\varepsilon, \quad \operatorname{dist}(0, \{u < 0\}) \leq \sigma_\varepsilon,$

then there exits a two-half-space global solution $u_0 \in P_\infty(M)$ such that

$$\|u - u_0\|_{C^1(B_1)} \leq \varepsilon.$$

Proof. As before, fix $\varepsilon > 0$ and assume that the statement is false. Then we can find sequences $R_j \to \infty$, $\sigma_j \to 0$, and solutions u_j of Problem **C** in B_{R_j} with $|D^2 u_j| \leq M$ and

$$|\nabla u_j(0)| \leq \sigma_j, \quad \operatorname{dist}(0, \{u_j > 0\}) \leq \sigma_j, \quad \operatorname{dist}(0, \{u_j < 0\}) \leq \sigma_j,$$

but such that

$$\|u_j - u_0\|_{C^1(B_1)} > \varepsilon, \quad \text{for any } u_0 \in P_\infty(M).$$

Noticing that we have the estimates

$$|\nabla u_j(x)| \leq \sigma_j + M|x|, \quad |x| < R_j,$$

and

$$|u_j(0)| \leq \sigma_j(\sigma_j + M\sigma_j)$$

we may assume that $u_j \to \hat{u}_0 \in P_\infty(M)$ in $C^1_{\mathrm{loc}}(\mathbb{R}^n)$. We claim that 0 is a branch point for $\hat{u}_0$. Indeed,

$$|\nabla \hat{u}_0(0)| = \lim_{j\to\infty} |\nabla u_j(0)| = 0, \quad \hat{u}_0(0) = 0.$$

Also, it is not hard to see that $0 \in \partial\{\hat{u}_0 > 0\} \cap \partial\{\hat{u}_0 < 0\}$ by using the nondegeneracy property. Hence, 0 is a branch point. But then Theorem 5.8 implies that $\hat{u}_0$ is a two-half-space solution and we arrive at a contradiction, since

$$\|u_j - \hat{u}_0\|_{C^1(B_1)} < \varepsilon, \quad \text{for } j \geq j_\varepsilon.$$

□

Notes

The original proof of the convexity of global solutions in the classical obstacle problem was given by Caffarelli [**Caf77, Caf80**] based on the "almost convexity" estimates of local solutions, as outlined in Exercise 5.1.

In dimension two, a complete classification of global solutions of the classical obstacle problem was obtained by Sakai [**Sak81**]. Namely, the coincidence set of global solutions is the limit of a family of ellipses: a line segment (finite or infinite), a half-plane, an ellipse (including a circle), or a parabola. The complements of the coincidence sets are also known as null quadrature domain.

In higher dimensions, there is a conjecture due to Shahgholian [**Sha92**], saying that the coincidence set of any global solution is the limit for a family of ellipsoids. It is also proved in that paper that such sets give rise to global solutions. Several attempts were made to fully classify global solutions, but the theory is still incomplete. We refer to DiBenedetto-Friedman [**DF86**], Friedman-Sakai [**FS86**], Ou [**Ou94**], Karp [**Kar94**], Karp-Margulis [**KM96, KM10**], and the references therein for further study on the topic.

The classification of global solutions in Problem **A** is due to Caffarelli-Karp-Shahgholian [**CKS00**], and in Problem **B** it is due to Caffarelli-Salazar-Shahgholian [**CSS04**]. The one in Problem **C** is obtained by Shahgholian-Uraltseva-Weiss [**SUW04**]. The methods in all three papers are heavily based on the ACF monotonicity formula (or the CJK estimate) combined with Weiss's monotonicity formula (or that of Spruck [**Spr83**] in the original treatment of Problem **A**).

The approximation by global solutions is one of the central ideas of the blowup method of Caffarelli [**Caf80**].

The classification of global solutions in the parabolic case can be found in Caffarelli-Petrosyan-Shahgholian [**CPS04**] for Problem **A** and Shahgholian-Uraltseva-Weiss [**SUW09**] for Problem **C**.

Exercises

5.1. This exercise gives an alternative proof of Theorem 5.1.

(i) Let $\phi \in C^{1,1}([0,h^{1/2}])$ be such that

$$|\phi(0)| \le C_n h^2, \quad |\phi'(0)| \le C_n h, \quad \phi(h^{1/2}) \ge 0.$$

Then, there exists $t \in (0,h^{1/2})$ such that $\phi''(t) \ge -C_n h^{1/2}$.

(ii) Let $u \ge 0$ be a solution of the classical obstacle problem $\Delta u = \chi_{\{u>0\}}$ and suppose that $B_r(x^0) \subset \Omega(u)$ is tangent to $\Gamma(u)$ at y^0.

Fix a unit vector e and let $\alpha \geq 0$ be such that

$$\partial_{ee} u \geq -\alpha \quad \text{in } B_r(x^0).$$

Then there exist positive dimensional constants C and M such that

$$\partial_{ee} u \geq -\alpha + C\alpha^M \quad \text{in } B_{r/2}(x^0).$$

Outline of the proof: Assume $r = 1$ and $y^0 = 0$ and consider the point $y^1 = hx^0$. Then we can apply (i) to find $z \in [y^1, y^1 + h^{1/2}e]$ such that

$$\partial_{ee} u(z) \geq -C_0 h^{1/2}.$$

If we assume that $x^0 \cdot e \geq 0$ (otherwise replace e by $-e$), then $z \in B_{1-h}(x^0)$. Next, choose $h = (\alpha/2C_0)^2$, so that $C_0 h^{1/2} = \alpha/2$. Then, consider the function $w = \partial_{ee} u + \alpha$ in $B_1(x^0)$. We have

$$\Delta w = 0 \quad \text{in } B_1(x^0), \qquad w \geq 0 \quad \text{in } B_1(x^0), \qquad \sup_{B_{1-h}(x^0)} w \geq \alpha/2.$$

Applying the Harnack inequality, we obtain that

$$\inf_{B_{1-h}(x^0)} w \geq C_n \alpha h^n,$$

which implies the required statement.

(iii) Let $u \geq 0$ be a solution of the classical obstacle problem $\Delta u = \chi_{\{u>0\}}$ in B_1. Let e be any unit vector and let $\alpha \geq 0$ be such that

$$\partial_{ee} u \geq -\alpha \quad \text{in } B_1 \cap \Omega(u).$$

Then there exist positive dimensional constants C and M such that

$$\partial_{ee} u \geq -\alpha + C\alpha^M \quad \text{in } B_{1/2} \cap \Omega(u).$$

Hint: For $x^0 \in B_{1/2} \cap \Omega(u)$ apply (ii) with $r = \operatorname{dist}(x^0, \Gamma(u))$.

(iv) Prove the following theorem: Let $u \geq 0$ be a solution of the classical obstacle problem $\Delta u = \chi_{\{u>0\}}$ in B_1. Then there exists a modulus of continuity $\sigma(r)$ depending only on the dimension, so that

$$\inf_{B_r} \partial_{ee} u \geq -\sigma(r), \quad 0 < r < 1,$$

for any direction e. (In fact, one can take $\sigma(r) = (\log 1/r)^{-\varepsilon}$ for some small $\varepsilon > 0$.)

(v) Deduce Theorem 5.1 from (iv).

5.2. Let $u \geq 0$ be a global solution of the classical obstacle problem. Show that u has at most quadratic growth at infinity, i.e. $u(x) \leq C(|x|^2 + 1)$. (This is generally not true if u solves Problem **A**.)

5.3. Give an alternative proof of Proposition 5.3 for Problem **A** by using Weiss's monotonicity formula. Also prove that if a blowup of a global solution u at x^0 is polynomial, then u itself is a polynomial.

5.4. If $u \in C^1(\mathbb{R}^n_+)$ and $\partial_e u$ does not change sign in $\mathbb{R}^n_+$ for any direction $e \in \partial\mathbb{R}^n_+$, then u is two-dimensional.

5.5. If $v_k \in C(\overline{B}_1)$, $v_k \to v_0$ in $C(\overline{B}_1)$ and

$$\min \operatorname{diam}\{v_k = 0\} \geq \sigma > 0,$$

then

$$\min \operatorname{diam}\{v_0 = 0\} \geq \sigma > 0.$$

5.6. If $u \in C^1(\mathbb{R}^n)$ is a convex function and $a = \inf_{\mathbf{R}^n} u$, then

$$\{u = a\} = \{|\nabla u| = 0\}.$$

5.7. If u is a global solution, then its shrinkdown at $x^0 \in \mathbb{R}^n$ does not depend on x^0.

5.8. Prove that the only global solution of the classical obstacle problem $\Delta u = \chi_{\{u>0\}}$, $u \geq 0$ in $\mathbb{R}^n$, satisfying $\sup_{\mathbb{R}^n} |D^2 u| < \infty$ and vanishing on a half-space has the form $u(x) = \frac{1}{2}(x \cdot e - c)_+^2$ with $|e| = 1$, $c \in \mathbb{R}$.

5.9. Prove the following generalization of Theorem 5.8, removing the assumption that $|\nabla u(0)| = 0$. If u is a solution of Problem **C** in $\mathbb{R}^n$ with $\|D^2 u\|_{L^\infty(\mathbb{R}^n)} \leq M$ and $0 \in \Gamma_+(u) \cap \Gamma_-(u)$, then

$$u(x) = \frac{\lambda_+}{2}[(x \cdot e)^+]^2 - \frac{\lambda_-}{2}[(x \cdot e)^-]^2 + a(x \cdot e),$$

with $a \geq 0$.

Chapter 6

Regularity of the free boundary: uniform results

In §6.1 we establish the uniform Lipschitz regularity of the free boundary in our model problems in terms of geometric conditions satisfied by the solution. The main difference from the results in Chapter 4 is that the estimates are uniform across a class of solutions rather than for specific solutions. Further, in §6.2, we show the $C^{1,\alpha}$ regularity of the free boundary in terms of the Lipschitz norm, by an application of the boundary Harnack principle. Next, in §6.3 we prove uniform C^1 regularity of the free boundaries near branch points in Problem **C**, and in §6.3.2 we describe a counterexample that the free boundaries are generally not $C^{1,\mathrm{Dini}}$. Finally, in §6.4, we use the partial hodograph-Legendre transformation to prove the real analyticity of the free boundaries near regular points in Problems **A** and **B**.

6.1. Lipschitz regularity of the free boundary

6.1.1. Problem A. We have shown in §5.4 that any local solutions $u \in P_R(M)$, for large enough R, can be approximated by a convex global solution u_0, provided $\Omega^c(u) \cap B_1$ is sufficiently thick. Moreover, the approximating global solution u_0 will contain a ball in $\Omega^c(u_0) \cap B_1$. Since the latter set is convex, this will immediately imply the Lipschitz regularity of $\Gamma(u_0)$. To give a more accurate version of this statement, we introduce the following family of rectangles:

$$K(\delta, s, h) := \{|x'| < \delta, -s \leq x_n \leq h\}, \quad \text{for } \delta, s, h > 0.$$

Recall also that we denote

$$\mathcal{C}_\delta = \{x = (x', x_n) \in \mathbb{R}^n : x_n > \delta |x'|\}, \quad \text{for } \delta > 0.$$

Lemma 6.1. *Let $u \in P_\infty(M)$ be a convex global solution of Problem* **A** *such that $\Omega^c(u) \cap B_1$ contains a ball $B = B_\rho(-se_n)$, for some $0 < \rho < s \le 1 - \rho$. Then the following assertions hold:*

(i) *For any unit vector $e \in \mathcal{C}_{4/\rho}$ we have*

$$\partial_e u \ge 0 \quad \text{in } K(\rho/2, s, 1).$$

(ii) *The portion $\Gamma \cap K(\rho/8, s, 1/2)$ of the free boundary is a Lipschitz graph*

$$x_n = f(x'),$$

where f is concave in x' and

$$|\nabla_{x'} f| \le \frac{4}{\rho}.$$

(iii) *There exists a constant $C_0 = C_0(\rho, M, n) > 0$ such that*

$$C_0 \partial_e u - u \ge 0 \quad \text{in } K(\rho/16, s, 1/2),$$

for any $e \in \mathcal{C}_{8/\rho}$.

Proof. Let $e \in \mathcal{C}_{4/\rho}$ and recall that $u \ge 0$ in $\mathbb{R}^n$. Then observe the following geometric property: every ray emanating from a point in $K(\rho/2, s, 1)$ in the direction $-e$ intersects the ball $B = B_\rho(-se_n)$. Since $\partial_e u = 0$ on B and $\partial_{ee} u \ge 0$ in $\mathbb{R}^n$ (from convexity), we readily obtain (i).

Further, it is easy to see that for $x^0 \in \Gamma$,

$$(x^0 + \mathcal{C}_{4/\rho}) \cap K(\rho/2, s, 1) \subset \{u > 0\}, \tag{6.1}$$

$$(x^0 - \mathcal{C}_{4/\rho}) \cap K(\rho/2, s, 1) \subset \{u = 0\}. \tag{6.2}$$

Then, using (i) we find the representation $x_n = f(x', t)$ in $K(\rho/8, s, 1/2)$, with the Lipschitz estimate $|\nabla_{x'} f| \le 4/\rho$.

To prove (iii), we first show that

$$\partial_{e_n} u(x) \ge C\rho^2 |x_n - f(x')| \quad \text{in } \Omega \cap K(\rho/8, s, 1/2), \tag{6.3}$$

for a dimensional constant C. To this end, fix $x \in \Omega \cap K(\rho/8, s, 1/2)$. By nondegeneracy (see Lemma 3.1) there exists $y \in B_{(\rho/8)|x_n - f(x')|}(x)$ such that

$$u(y) \ge C\rho^2 |x_n - f(x')|^2.$$

Note that a larger ball $B_{(\rho/4)|x_n - f(x')|}(x)$ is contained in $((x', f(x')) + \mathcal{C}_{4/\rho}) \cap K(\rho/2, s, 1)$. By the convexity argument in the beginning of the proof, $\partial_{e_n} u$ is increasing in the e_n-direction, and therefore we obtain that

$$\partial_{e_n} u(y) \ge \frac{u(y)}{|y_n - f(y')|} \ge \frac{C\rho^2 |x_n - f(x')|^2}{(3/2 + \rho/8)|x_n - f(x')|} \ge C\rho^2 |x_n - f(x')|.$$

The inequality (6.3) follows now from the Harnack inequality applied to the positive harmonic function $\partial_{e_n} u$ in the ball $B_{(\rho/4)|x_n - f(x')|}(x)$.

On the other hand, using that $u = \partial_{e_n} u = 0$ at $(x', f(x'))$ and that $\partial_{e_n e_n} u \leq M$, and integrating, we readily have

$$u(x) \leq \frac{1}{2} M |x_n - f(x')|^2.$$

Combining this with (6.3), we obtain the statement of part (iii) for $e = e_n$. To complete the proof we now observe that if $e \in \mathcal{C}_{8/\rho}$ and we make a rotation of coordinate axes that takes e to e_n, then the conclusion of part (ii) as well as the proof of (iii) for $e = e_n$ will go through with $\mathcal{C}_{4/\rho}$ replaced by a narrower cone $\mathcal{C}_{8/\rho}$ and $K(\rho/8, s, 1/2)$ by $K(\rho/16, s, 1/2)$.

It is also noteworthy that it is possible to prove part (iii) by using a compactness argument (see Exercise 6.1). □

Now, we prove the Lipschitz regularity of the free boundary for solutions in large balls.

Theorem 6.2. *For every $\sigma \in (0, 1]$ there exists $R_\sigma = R_\sigma(M, n)$ such that if $u \in P_{R_\sigma}(M)$ is a solution of Problem* **A** *and $\delta(1, u) \geq \sigma$, then $\Gamma(u) \cap B_{c_n \sigma}$ is Lipschitz regular with a Lipschitz constant $L = L(\sigma, n, M)$.*

Proof. Fix a small $\varepsilon = \varepsilon(\sigma, M, n) > 0$, and apply Lemma 5.13. So, if $R > R_{\varepsilon,\sigma}$, we can find a convex global solution u_0 such that conditions (i)–(iii) in Lemma 5.13 are satisfied. In particular, $\Omega^c(u_0) \cap B_1$ must contain a ball B of radius $\rho = \sigma/2n$. Without loss of generality we may assume that $B = B_\rho(-se_n)$ for some $\rho \leq s \leq 1 - \rho$ and we will also have that $B_{\rho/2}(-se_n) \subset \Omega^c(u) \cap B_1$. We now apply Lemma 6.1, which gives that

$$C_0 \partial_e u_0 - u_0 \geq 0 \quad \text{in } K(\rho/16, s, 1/2),$$

for any $e \in \mathcal{C}_{8/\rho}$.

Further, if ε is small enough, the approximation $\|u - u_0\|_{C^1(B_1)} \leq \varepsilon$ implies that

$$C_0 \partial_e u - u \geq -(C_0 + 1)\varepsilon > -(\rho/16)^2/8n \quad \text{in } K(\rho/16, s, 1/2).$$

Recalling Lemma 4.3, we obtain

$$C_0 \partial_e u - u \geq 0 \quad \text{in } K(\rho/32, s, 1/4).$$

(To be more accurate, one needs to apply Lemma 4.3 in every ball $B_{\rho/16}(te_n)$ for $t \in [-s, 1/4]$.) The latter inequality can be rewritten as

$$\partial_e (e^{-C_0 (x \cdot e)} u) \geq 0 \quad \text{in } K(\rho/32, s, 1/4).$$

Taking $e = e_n$ and noting that $u = 0$ on $B'_{\rho/32} \times \{-s\}$, we obtain after integration that

$$u \geq 0 \quad \text{in } K(\rho/32, s, 1/4).$$

Combined with the previous inequality this gives

$$\partial_e u \geq 0 \quad \text{in } K(\rho/32, s, 1/4),$$

for any $e \in \mathcal{C}_{8/\rho}$.

The rest of the proof is now left to the reader as an exercise. □

Next, we give a reformulation of Theorem 6.2.

Theorem 6.3 (Thickness implies Lipschitz: Problem **A**). *There exists a modulus of continuity $\sigma(r) = \sigma_{M,n}(r)$ such that if $u \in P_1(M)$ is a solution of Problem* **A** *and $\delta(r, u) \geq \sigma(r)$ for some value $r = r_0 \in (0, 1)$, then $\Gamma \cap B_{c_n r_0 \sigma(r_0)}$ is a Lipschitz graph with a Lipschitz constant $L \leq L(n, M, r_0)$.*

Proof. This is basically a rescaled version of Theorem 6.2.

Note that in Theorem 6.2 one can take the function $\sigma \mapsto R_\sigma$ to be monotone and continuous in σ and such that $\lim_{\sigma\to 0+} R_\sigma = \infty$. Then let $r \mapsto \sigma(r)$ be the inverse of the mapping $\sigma \mapsto 1/R_\sigma$ so that we have

$$R_{\sigma(r)} = 1/r.$$

Now, if $\delta(r_0, u) \geq \sigma(r_0)$, then the rescaling

$$u_{r_0}(x) = \frac{u(r_0 x)}{r_0^2} \in P_{1/r_0}(M)$$

satisfies

$$\delta(1, u_{r_0}) \geq \sigma(r_0).$$

Because of the identity $R_{\sigma(r_0)} = 1/r_0$ we can apply Theorem 6.2. Then scaling back to u we obtain the corresponding statement for the free boundary of u. □

6.1.2. Problem B.

We are going to show here that Theorems 6.2 and 6.3 hold also for solutions of Problem **B**.

Lemma 6.4. *Let u be as in Lemma* 6.1. *Then we also have*

$$C_0 \partial_e u - |\nabla u|^2 \geq 0 \quad \text{in } K(\rho/16, s, 1/2)$$

for any $e \in \mathcal{C}_{8/\rho}$, where $C_0 = C_0(\rho, M, n)$.

Proof. The proof is completely analogous to that of Lemma 6.1 (iii). □

Theorem 6.5 (Thickness implies Lipschitz: Problem **B**). *Theorem* 6.2 *holds also for solutions of Problem* **B**.

Proof. Arguing similarly to the case of Problem **A**, but using Lemma 4.13 instead of Lemma 4.3 and Lemma 6.4 instead of Lemma 6.1 (iii), we can show that

$$C_0\partial_e u - |\nabla u|^2 \geq 0 \quad \text{in } K(\rho/32, s, 1/4),$$

for any $e \in \mathcal{C}_{8/\rho}$, which immediately implies that

$$\partial_e u \geq 0 \quad \text{in } K(\rho/32, s, 1/4).$$

We next claim that u is constant on $\Omega^c(u) \cap K(\rho/32, s, 1/4)$. The argument is similar to the one in the proof of Theorem 4.11. It will suffice to show that for every point $x = (x', x_n) \in \Omega^c$ the segment joining $(x', -s)$ and x is completely contained in Ω^c. Indeed, if the latter statement is false, we can find in $K(\rho/32, s, 1/4)$ two points $x = (x', x_n) \in \Omega^c$ and $\tilde{x} = (x', \tilde{x}_n) \in \Omega$ such that $\tilde{x}_n < x_n$. Without loss of generality we may assume that $x \in \Gamma$. Now, let us take a small ball $B_\varepsilon(\tilde{x})$ and start moving this ball from $\tilde{x}$ to x along the x_n axis, reducing its radius proportionally to the distance from x. Stop moving if the ball touches the free boundary Γ at some point. Call this point ζ^0. If the moving ball does not touch Γ, then let $\zeta^0 = x$. In either case there will exist a cone $\mathcal{C}$ with axis e_n and $\eta > 0$ such that

$$\zeta^0 - (\mathcal{C} \cap B_\eta) \subset \Omega(u).$$

Consider now a blowup $\hat{u}_0$ of u at ζ^0. Since $\partial_e u \geq 0$ in $K(\rho/32, s, 1/4)$ for e in a cone of directions $\mathcal{C}_{8/\rho}$, $\hat{u}_0$ must necessarily be a half-space solution $\hat{u}_0(x) = \frac{1}{2}(x \cdot e_0)_+^2$ with e_0 satisfying $e \cdot e_0 \geq 0$ for all $e \in \mathcal{C}_{4/\rho}$. On the other hand, one must also have

$$-\mathcal{C} \subset \Omega(\hat{u}_0),$$

which implies that $e_0 \cdot e_n < 0$, a contradiction. Thus, we obtain that the line segment joining x and $(x', -s)$ is fully contained in Ω^c for $x \in \Omega^c(u) \cap K(\rho/16, s, 1/4)$. Consequently, u is constant in $\Omega^c(u) \cap K(\rho/16, s, 1/4)$. Subtracting a constant, we reduce the problem to the case of Problem **A**. □

As an immediate corollary we obtain the following statement.

Theorem 6.6. *Theorem 6.3 holds also for solutions of Problem* **B**. □

6.1.3. Problem C. Next we use an approximation argument with global solutions to deduce the following theorem for solutions of Problem **C**. Similarly to Lemma 5.15, the result is stated for a class of functions that is slightly wider than $P_1(M)$, which will be needed for the proof of the C^1 regularity of the free boundaries (even for functions in $P_1(M)$).

Theorem 6.7. *For any $L > 0$ there exist positive σ_0 and r_0 depending only on $\lambda_\pm$, M, n, and L such that if u is a solution of Problem* **C** *in B_1 with $\|D^2u\|_{L^\infty(B_1)} \leq M$ and*

$$|\nabla u(0)| \leq \sigma_0, \quad \{\pm u > 0\} \cap B_{\sigma_0} \neq \emptyset, \tag{6.4}$$

then $\Gamma_\pm(u) \cap B_{r_0}$ are Lipschitz graphs with a Lipschitz constant not greater than L.

Proof. Fix small $\varepsilon > 0$, to be specified later, and let σ_ε and $R_\varepsilon > 1$ be as in Lemma 5.15. Put $\sigma_0 = \sigma_\varepsilon / R_\varepsilon$. Now, if u satisfies (6.4), the rescaling

$$u_{1/R_\varepsilon}(x) = R_\varepsilon^2 u(x/R_\varepsilon)$$

satisfies the conditions of the approximation Lemma 5.15. Hence, there exists a two-half-space solution u_0 such that

$$\|u_{1/R_\varepsilon} - u_0\|_{C^1(B_1)} \leq \varepsilon.$$

Without loss of generality we may assume that $u_0(x) = \frac{\lambda_+}{2}(x_n^+)^2 - \frac{\lambda_-}{2}(x_n^-)^2$. Then we have

$$C\partial_e u_0 - |u_0| \geq 0 \quad \text{in } B_1, \quad C = \frac{1}{2}\sqrt{1+L^{-2}},$$

for any $e \in \mathcal{C}_L$. From the approximation we have

$$C\partial_e u_{1/R_\varepsilon} - |u_{1/R_\varepsilon}| \geq -(C+1)\varepsilon > -\lambda_{\min}/8n \quad \text{in } B_1,$$

provided $\varepsilon = \varepsilon(\lambda_\pm, M, m, L) > 0$ is small enough. Then Lemma 4.23 implies

$$C\partial_e u_{1/R_\varepsilon} - |u_{1/R_\varepsilon}| \geq 0 \quad \text{in } B_{1/2},$$

and consequently

$$\partial_e u_{1/R_\varepsilon} \geq 0 \quad \text{in } B_{1/2}$$

for any $e \in \mathcal{C}_L$. From this, arguing as in Section 4.4, we conclude that $\Gamma_\pm(u_{1/R-\varepsilon}) \cap B_{1/2}$ are Lipschitz graphs with a constant L. Scaling back, we obtain the Lipschitz regularity of $\Gamma_\pm(u) \cap B_{1/2R_e}$. Thus, we can take $r_0 = 1/2R_\varepsilon$. □

6.2. $C^{1,\alpha}$ Regularity of the free boundary: Problems A and B

In this section we give a proof of the $C^{1,\alpha}$ regularity of the free boundary. We follow the original idea of Athanasopoulos-Caffarelli [**AC85**] based on the boundary Harnack principle (also known as the local comparison theorem) for Lipschitz domains.

6.2.1. Boundary Harnack principle.

Theorem 6.8 (Boundary Harnack principle). *Let $f : \mathbb{R}^{n-1} \to \mathbb{R}$ be a Lipschitz function with $f(0) = 0$ and a Lipschitz constant M, and let*

$$\Omega = \{(x', x_n) : x_n > f(x')\}.$$

Let u_1 and u_2 be two positive harmonic functions in $B_R \cap \Omega$, continuously vanishing on $B_R \cap \partial\Omega$. Then

$$\sup_{B_{R/K}\cap\Omega} \frac{u_1}{u_2} \le C \inf_{B_{R/K}\cap\Omega} \frac{u_1}{u_2}, \tag{6.5}$$

where $C, K > 1$ depend only on M and n.

Moreover, there exist constants $C > 0$ and $0 < \alpha < 1$, depending only on M and n, such that

$$\operatorname*{osc}_{B_\rho\cap\Omega} \frac{u_1}{u_2} \le C \left(\frac{\rho}{R}\right)^\alpha \operatorname*{osc}_{B_R\cap\Omega} \frac{u_1}{u_2}, \tag{6.6}$$

for any $0 < \rho < R$. □

We will not discuss the detailed proof of the theorem, which can be found for instance in [**CS05**, Chapter 11]. However, we will make the following series of remarks.

1) Instead of being harmonic, one may ask u_k, $k = 1, 2$, to be (weak) solutions of a uniformly elliptic equation in divergence form with bounded measurable coefficients, i.e.

$$\begin{aligned} \operatorname{div}(A(x)\nabla u_k) &= 0 \quad \text{in } B_R \cap \Omega, \\ u_k &= 0 \quad \text{on } B_R \cap \partial\Omega, \end{aligned}$$

where $A(x) = (a_{ij}(x))$ is an $n \times n$ symmetric matrix-valued function with bounded measurable entries and such that for some $\lambda > 0$,

$$\lambda|\xi|^2 \le \xi \cdot A(x)\xi \le \frac{1}{\lambda}|\xi|^2 \quad \text{for any } \xi \in \mathbb{R}^n.$$

Then the constants in (6.5) and (6.6) will depend additionally on λ.

2) In fact, the class of equations in divergence form is more natural for the boundary Harnack principle, because of the invariance under bi-Lipschitz transformations. Thus, if we consider the transformation $y = T(x)$ given by

$$y_i = x_i, \quad i = 1, \dots, n-1, \quad y_n = x_n - f(x'),$$

then it will map Ω to the half-space $\mathbb{R}^n_+ = \{y_n > 0\}$ (it will straighten the boundary of Ω, so to say) and will transform u_k to $v_k(y) = u_k(T^{-1}(y))$, $k = 1, 2$, which satisfy

$$\begin{aligned} \operatorname{div}(B(y)\nabla_y v_k) &= 0 \quad \text{in } B^+_{R/K} \subset T(B_R \cap \Omega), \\ v_k &= 0 \quad \text{on } B'_{R/K}, \end{aligned}$$

where $B(y) = \det(DT(x))^{-1} DT(x)^t A(x) DT(x)$, and $K = K(M, n)$. So Theorem 6.8 will follow once we have it for the pair v_1, v_2 as above.

3) For a domain Ω as in Theorem 6.8 and for $x^0 \in \partial\Omega$ let

$$a_r(x^0) = x^0 + \frac{r}{2} e_n,$$

which is known as a *corkscrew* point and has the property that

$$\frac{1}{Cr} < \operatorname{dist}(a_r(x^0), \Omega), \quad |a_r(x^0) - x^0| < Cr,$$

for $C = C(M,n) > 0$. One of the steps in the proof of Theorem 6.8 is the inequality

$$\sup_{B_{r/K}\cap\Omega} \frac{u_1}{u_2} \leq C\frac{u_1(a_r(0))}{u_2(a_r(0))}.$$

In particular, we note that in (6.6) one can replace

$$\operatorname*{osc}_{B_R\cap\Omega} \frac{u_1}{u_2} \quad \text{by} \quad \frac{u_1(a_r(0))}{u_2(a_r(0))}.$$

4) Finally, we note that Theorem 6.8 is trivially generalized to the Lipschitz domains Ω, i.e. the domains such that for any $x^0 \in \partial\Omega$,

$$B_{r_0}(x^0) \cap \Omega = B_{r_0}(x^0) \cap \{x_n > f(x')\},$$

for some fixed $r_0 > 0$ and a certain Lipschitz f with a fixed Lipschitz constant M. Moreover, Theorem 6.8 can be generalized to a wider class of domains, known as NTA (nontangentially accessible) domains (see Jerison-Kenig [**JK82**] and Exercise 6.3), or even more exotic uniformly John domains (see Aikawa-Lundh-Muzutani [**ALM03**]).

6.2.2. Lipschitz implies $C^{1,\alpha}$: Problems A, B. We now apply the boundary Harnack principle to bootstrap the Lipschitz regularity of the free boundary to the $C^{1,\alpha}$ regularity in Problems **A** and **B**.

Theorem 6.9. *For every $\sigma > 0$ there exists $R_\sigma = R_\sigma(M,n)$ such that if $u \in P_R(M)$ is a solution of Problem* **A** *or* **B** *with $\delta(1,u) \geq \sigma$, for $R > R_\sigma$, then $\Gamma(u) \cap B_{c_n\sigma}$ is a $C^{1,\alpha}$ graph with $\alpha = \alpha_{\sigma,M,n} \in (0,1)$ and the $C^{1,\alpha}$-norm $C \leq C(\sigma, n, M)$.*

Proof. Let R_σ be as in Theorem 6.2 and suppose the ball $B = B_\rho(-se_n)$ is as in Lemma 6.1 (with $\rho = \sigma/2n$) so that u vanishes on $B_{\rho/2}(-se_n)$. Then, as it follows from the proof of Theorem 6.2, we have

$$\partial_e u \geq 0 \quad \text{in } K(\rho/32, s, 1/4),$$

for any unit vector $e \in \mathcal{C}_{8/\rho}$. Consider now two functions of the above type,

$$\begin{aligned} u_1 &= \partial_e u, \\ u_2 &= \partial_{e_n} u, \end{aligned}$$

with e sufficiently close to e_n. Since we already know that $\Omega \cap K(\rho/32, s, 1/4)$ is given as an epigraph $x_n > f(x')$ of a Lipschitz function, the boundary Harnack principle implies that the ratio

$$\frac{u_1}{u_2}$$

is C^α regular in $\Omega\cap K(\rho/64,s,1/8)$ up to Γ, with $0<\alpha<1$. Moreover, the C^α-norm depends on ρ, n, M, the Lipschitz norm of Γ, as well as on the lower bound for

$$u_2(a_{\rho/32}(0)),\quad a_{\rho/32}(0)=\frac{\rho}{64}e_n;$$

see remark 3) after Theorem 6.8. We claim that

$$u_2(a_{\rho/32}(0))=\partial_{e_n}u(a_{\rho/32}(0))\geq c_0(\rho,n,M)>0.$$

In fact, we have already shown this in (6.3). Hence,

$$\frac{\partial_e u}{\partial_{e_n}u}$$

is C^α up to Γ in $\Omega\cap K(\rho/64,s,1/8)$. Now taking $e=(e_n+\varepsilon e_j)/\sqrt{1+\varepsilon^2}$ for a small $\varepsilon\leq\varepsilon(\rho,n,M)$ and $j=1,\dots,n-1$, we obtain that the ratios

$$\frac{\partial_{e_j}u}{\partial_{e_n}u},\quad j=1,\dots,n-1,$$

are C^α. We claim that this implies that $\Gamma\cap K(\rho/64,s,1/8)$ is a graph $x_n=f(x')$ with

$$\|f\|_{C^{1,\alpha}(B'_{\rho/64})}\leq C(\rho,n,M).$$

Indeed, first notice that the level sets $\{u=\eta\}\cap B_{\rho/64}$ with a small $\eta>0$ are Lipschitz graphs $x_n=f_\eta(x')$, which follows from the positivity of $\partial_e u$ for $e\in\mathcal{C}_{8/\rho}$. Next, notice that on these level sets we have

$$\partial_{e_j}f_\eta=\frac{\partial_{e_j}u}{\partial_{e_n}u},\quad j=1,\dots,n-1,$$

which implies that the functions f_η are $C^{1,\alpha}$ on $B'_{\rho/64}$ uniformly for $0<\eta<\eta(\rho,n,M)$. On the other hand, it is easy to see that

$$f(x')=\lim_{\eta\to0+}f_\eta(x'),\quad\text{for any } x'\in B'_{\rho/64},$$

which implies that f is also $C^{1,\alpha}$ on $B'_{\rho/64}$. □

The rescaled version of the previous theorem is as follows.

Theorem 6.10. *There exists a modulus of continuity $\sigma(r)=\sigma_{M,n}(r)$ such that if $u\in P_1(M)$ and $\delta(r,u)\geq\sigma(r)$ for some $r=r_0\in(0,1)$, then $\Gamma(u)\cap B_{c_nr_0\sigma(r_0)}$ is a $C^{1,\alpha}$-graph with $\alpha=\alpha_{r_0,M,n}\in(0,1)$ and the $C^{1,\alpha}$-norm $C\leq C(r_0,M,n)$.*

Analyzing the above proof, we realize that we have actually proved the following result.

Theorem 6.11 (Lipschitz implies $C^{1,\alpha}$). *Let $u \in P_1(M)$ be a solution of Problem **A** or **B** such that*

$$\Gamma(u) = \{x_n = f(x')\} \cap B_1, \quad \Omega(u) = \{x_n > f(x')\} \cap B_1,$$

where f is Lipschitz continuous on B_1' with a Lipschitz constant L. Then

$$|\nabla f(x') - \nabla f(y')| \leq C|x' - y'|^\alpha, \quad x', y' \in B_\delta',$$

where $\delta > 0$, $C > 0$ and $0 < \alpha < 1$ depend only on n, M, and L.

6.3. C^1 regularity of the free boundary: Problem C

6.3.1. C^1 regularity.

Theorem 6.12. *Let u be a solution of Problem **C** in B_1 with $\|D^2 u\|_{L^\infty(B_1)} \leq M$. Then there are constants $\sigma_0 > 0$ and $r_0 > 0$ such that if*

$$|\nabla u(0)| \leq \sigma_0, \quad \Omega_\pm(u) \cap B_{\sigma_0} \neq \emptyset, \tag{6.7}$$

then $\Gamma_\pm(u) \cap B_{r_0}$ are C^1 surfaces. The constants σ_0, r_0 and the modulus of continuity of the normal vectors to these surfaces depend only on $\lambda_\pm$, M and the space dimension n.

Remark 6.13. The C^1 regularity is optimal in the sense that the graphs in general are not of class $C^{1,\mathrm{Dini}}$. This means that the normal to the free boundary might not be Dini continuous; i.e., if ω is the modulus of continuity of the normal vector, then it might be such that

$$\int_0^1 \frac{\omega(t)}{t} dt = \infty.$$

Corollary 6.14. *Let $u \in P_1(M)$ and suppose that $0 \in \Gamma^0(u)$ is a two-phase point. Then there is a constant $r_0 > 0$ such that $\Gamma_\pm(u) \cap B_{r_0}$ are C^1 surfaces. The constant r_0 and the modulus of continuity of the normal vectors to these surfaces depend only on $\lambda_\pm$, M, and the space dimension n.*

Proof of Theorem 6.12. From Theorem 6.7 we know that $\Gamma_\pm(u) \cap B_{r_0}$ are given as Lipschitz graphs (after a suitable rotation of coordinate axes)

$$x_n = f_\pm(x'),$$

with Lipschitz continuous $f_\pm$ satisfying $|\nabla_{x'} f_\pm(x')| \leq L < 1$, for $(x', f_\pm(x')) \in \Gamma_\pm(u) \cap B_{r_0}$. Moreover, we know that $f_\pm$ are differentiable and even C^1. So, it will suffice to show that the normals are equicontinuous on $\Gamma_\pm(u) \cap B_{r_0/2}$, for u in the class of solutions specified in the statement of the theorem.

We claim that for $\varepsilon > 0$ there is $\delta_\varepsilon > 0$ depending only on the parameters in the statement such that for any pair of free boundary points $y^1, y^2 \in \Gamma_+ \cap B_{r_0/2}$,

$$|y^1 - y^2| \leq \delta_\varepsilon \quad \Rightarrow \quad |\nu(y^1) - \nu(y^2)| \leq 2\varepsilon. \tag{6.8}$$

Fix $\varepsilon > 0$. Let σ_ε and r_ε denote the constants σ_0 and r_0 respectively in Theorem 6.7 for $L = \varepsilon$. In what follows $\rho_\varepsilon := \min\{r_\varepsilon, \sigma_\varepsilon\} r_0/4$.

Suppose first that u is nonnegative in $B_{\rho_\varepsilon}(y^1)$. Then we can apply the $C^{1,\alpha}$ regularity result to the scaled function

$$u_{y^1,\rho_\varepsilon}(x) := u(y^1 + \rho_\varepsilon x)/\rho_\varepsilon^2.$$

Since the $C^{1,\alpha}$-norm of the normal on $B_{c_0} \cap \partial\{u_{y^1,\rho_\varepsilon} > 0\}$ bounded by a constant C_0, where c_0 and C_0 depend only on the parameters in the statement, we may choose

$$\delta_\varepsilon := \min\{(\varepsilon/C_0)^{1/\alpha}, c_0\}\, \rho_\varepsilon$$

to obtain (6.8).

Next, suppose that u changes its sign in $B_{\rho_\varepsilon}(y^1)$. This means that $B_{\rho_\varepsilon}(y^1)$ intersects both $\{\pm u > 0\}$. If there is a point $y \in B_{\rho_\varepsilon}(y^1) \cap \Gamma_+$ such that $|\nabla u(y)| \le \rho_\varepsilon$, then the rescaling $u_{y,r_0/2}$ satisfies the conditions of Theorem 6.7 with $L = \varepsilon$. Namely,

$$|\nabla u_{y,r_0/2}(0)| \le \sigma_\varepsilon, \quad B_{\sigma_\varepsilon} \cap \{\pm u_{y,r_0/2} > 0\} \ne \emptyset.$$

Hence, the free boundary $\Gamma_+ \cap B_{r_\varepsilon r_0/2}(y) \supset \Gamma_+ \cap B_{\rho_\varepsilon}(y^1)$ is Lipschitz with Lipschitz norm not greater than ε. Hence (6.8) follows in this case with $\delta_\varepsilon := \rho_\varepsilon$.

Finally, if $|\nabla u| \ge \rho_\varepsilon$ for all points $y \in \Gamma_+ \cap B_{\rho_\varepsilon}(y^1)$, we proceed as follows. First note that since f_+ has a Lipschitz constant $L < 1$, we have $\partial_{e_n} u \ge \rho_\varepsilon/\sqrt{2}$ on $\Gamma_+ \cap B_{\rho_\varepsilon}(y^1)$. Also, by differentiating the identity $u(x', f_+(x')) = 0$, we obtain $\nabla_{x'} u + \partial_{e_n} u \nabla_{x'} f_+ = 0$ on $\Gamma_+ \cap B_{r_0/2}$. Therefore,

$$|\nu(y^1) - \nu(y^2)| \le \frac{C_0 M}{\rho_\varepsilon} |y^1 - y^2|,$$

for an absolute constant $C_0 > 0$, where M here is such that $|D^2 u| \le M$ in B_1. In particular, we may choose

$$\delta_\varepsilon := \frac{\varepsilon \rho_\varepsilon}{C_0 M}$$

to arrive at (6.8). □

6.3.2. Optimality of C^1 regularity. We now show the optimality of Theorem 6.12 by describing an explicit example from Shahgholian-Uraltseva-Weiss [**SUW07**] for which $\Gamma_\pm$ are not $C^{1,\mathrm{Dini}}$.

We will need some auxiliary results for our construction.

Lemma 6.15 (Nondegeneracy up to the boundary). *Let $v \ge 0$ be a solution of the obstacle problem*

$$\Delta v = \chi_{\{v>0\}} \quad \text{in } D$$

that vanishes on a subset Σ *of the boundary* ∂D. *Then, for any* $B_r(x_0) \subset \mathbb{R}^n$ *satisfying* $B_r(x_0) \cap \partial D \subset \Sigma$,

$$\sup_{D\cap B_r(x_0)} v \le r^2/(8n) \quad \Rightarrow \quad v \equiv 0 \quad \text{in } D \cap B_{r/2}(x_0)\,.$$

Proof. The proof follows from a comparison of v in $\{v > 0\} \cap B_{r/2}(y)$ with $w_y(x) = |x - y|^2/(2n)$ for $y \in B_{r/2}(x_0) \cap D$. □

Now let $\zeta \in C^\infty(\mathbb{R})$ be such that $\zeta = 0$ in $[-1/2, +\infty)$, $\zeta = 1/2$ in $(-\infty, -1]$ and ζ is strictly decreasing in $(-1, -1/2)$; see Fig. 6.1. Moreover, for $M \in [0, 1]$, define the function u_M as the solution of the classical obstacle problem (see Fig. 6.2):

$$(6.9) \qquad \begin{aligned} \Delta u_M &= \chi_{\{u_M>0\}} && \text{in } Q := \{x \in \mathbb{R}^2 : x_1 \in (0,1), x_2 \in (-1,0)\},\\ u_M(x_1,x_2) &= M\zeta(x_2) && \text{on } \{x_1 = 0\} \cap \partial Q,\\ u_M(x_1,x_2) &= M/2 && \text{on } \{x_1 = 1\} \cap \partial Q,\\ \partial_{x_2} u_M &= 0 && \text{on } (\{x_2 = -1\} \cup \{x_2 = 0\}) \cap \partial Q. \end{aligned}$$

The existence and uniqueness of such u_M is left as an exercise to the reader (see Exercise 6.4).

Moreover, the mapping $M \mapsto u_M$ is continuous from $[0, 1]$ to $C^{1,\alpha}(\overline{Q})$ as well as weakly continuous to $W^{2,p}(Q)$. Here we also note that $u_M = 0$ for $M = 0$ and that $u_1 > x_1^2/2$ by the comparison principle.

Next, we claim that $\partial_{x_2} u_M \le 0$ in Q. Indeed, suppose that $V = Q \cap \{\partial_{x_2} u_M > 0\}$ is nonempty. The derivative $\partial_{x_2} u_M$ is harmonic in V and nonpositive on ∂V, which implies by the maximum principle that $\partial_{x_2} u_M \le 0$ in V, a contradiction. Hence, $\partial_{x_2} u_M \le 0$ everywhere in Q. As a consequence, the free boundary of u_M is a graph of the x_1-variable.

Let us extend u_M by even reflection to $[0, 1] \times (0, 1)$ and keep the same notation u_M for the extended function. Now for every $M \in [0, 1]$ let $a_M \in [-1, 0]$ be such that

$$\partial(\operatorname{Int}\{u_M = 0\}) \cap (\{0\} \times [-1, 0]) = \{0\} \times [a_M, 0].$$

From the comparison principle it follows that the mapping $M \mapsto a_M$ is monotone. It is easy to see from Lemma 6.15 applied to the ball $B_{1/2}$ that

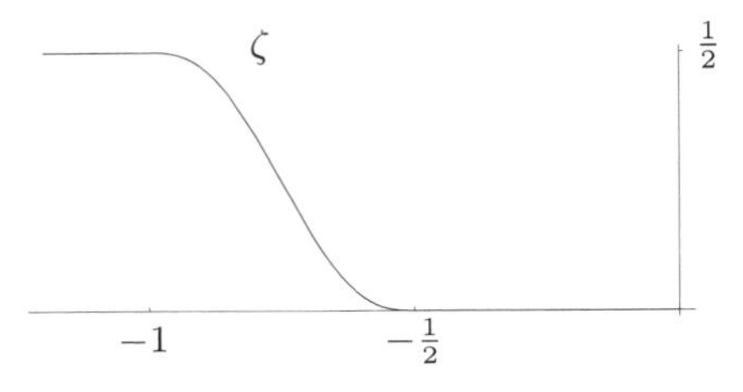

Figure 6.1. The graph of ζ

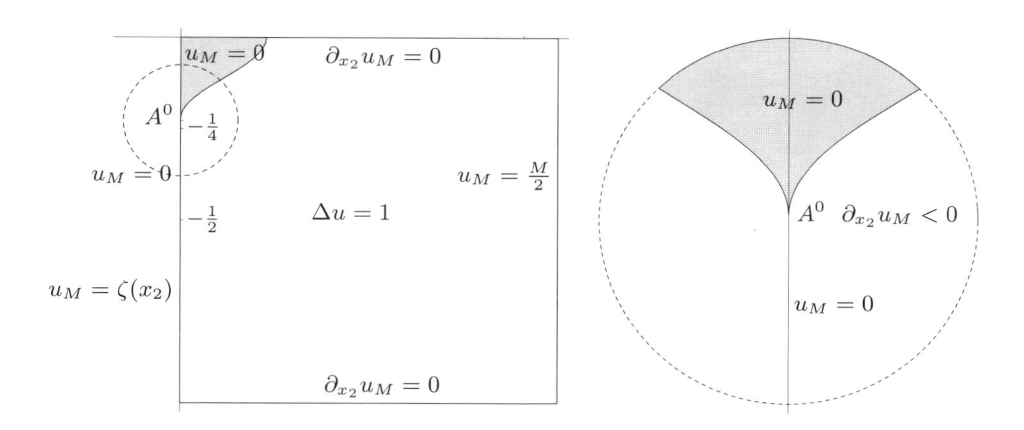

Figure 6.2. The function u_M

$u = 0$ in $Q \cap B_{1/4}$ for $M < 1/32$. Hence $a_M \leq -1/4$ for $M < 1/32$. We next claim that there exists $M_0 \in (0,1)$ such that $a_{M_0} \in [-1/4, 0)$.

Indeed, let $M_0 = \sup\{M \in [0,1] : a_M \leq -1/4\}$. The argument above implies that $M_0 \geq 1/32$. We next show that $M_0 < 1$. If, on the contrary, $a_M \leq -1/4$ for all $M \in (0,1)$, then letting $M \to 1$, we obtain $u_1 = |\nabla u_1| = 0$ on $\{0\} \times [-1/4, 0]$. Since also $\Delta u_1 = 1$ in Q, we obtain from the Cauchy-Kovalevskaya theorem (applied repeatedly to $w = u_1 - x_1^2/2$) that $u_1 \equiv x_1^2/2$ in Q; this is a contradiction in view of the boundary data of u_1. Thus, $M_0 \in (0,1)$.

Suppose now that $a_{M_0} < -1/4$. Then u_{M_0} vanishes in $B_\varepsilon(0, -1/4) \cap Q$ for a small $\varepsilon > 0$. From the continuous dependence of u_M on M and Lemma 6.15 it follows that if M is slightly above M_0, then u_M still vanishes on $B_{\varepsilon/2}(0, -1/4) \cap Q$ and therefore $a_M < -1/4$. However, the latter contradicts the choice of M_0. On the other hand, the argument above using the Cauchy-Kovalevskaya theorem can be applied again to justify that $a_{M_0} < 0$. Thus, we obtain that $a_{M_0} \in [-1/4, 0)$. Let A^0 denote the point $(0, a_{M_0})$.

Now we may extend u_{M_0} by odd reflection at the line $\{x_1 = 0\}$ to a solution u of Problem **C** in an open neighborhood of A^0 (see Fig. 6.2). Here $\lambda_+ = \lambda_- = 1$. The point A^0 is a branch point, so we may apply Theorem 6.12 to obtain that the free boundary is the union of two C^1-graphs in a neighborhood of A_0.

Suppose now, towards a contradiction, that $\partial\{u > 0\}$ is of class $C^{1,\mathrm{Dini}}$ in a neighborhood of A^0. Then by Widman [**Wid67**, Theorem 2.5], the Hopf principle holds at A^0 and tells us that

$$\limsup_{x_1 \to 0+} \frac{\partial_{x_2} u_{M_0}(x_1, a_{M_0})}{x_1} < 0.$$

But that contradicts Theorem 5.8, which shows that no blowup $\hat{u}_0$ of u_{M_0} at A^0 depends on x_2 and therefore

$$\limsup_{r\to 0+} \frac{\partial_{x_2} u_{M_0}(rx_1, a_{M_0})}{r} \geq \partial_{x_2}\hat{u}_0(x_1, 0) = 0.$$

Consequently $\partial\{u>0\}$ and $\partial\{u<0\}$ are not of class $C^{1,\mathrm{Dini}}$.

6.4. Higher regularity: Problems A and B

In this section we show how to use the so-called method of partial hodograph-Legendre transform to bootstrap the $C^{1,\alpha}$ regularity of the free boundary in Problems **A** and **B** to C^∞ and even real analyticity.

6.4.1. $C^{2,\alpha}$ regularity near regular points.

We start with a result that the $C^{1,\alpha}$ regularity of the free boundary Γ implies that $u \in C^{2,\alpha}(\Omega \cup \Gamma)$. Recall that we already know the C^2 regularity from Proposition 4.18; however, we also need a control over the modulus of continuity of D^2u in the application to the partial hodograph-Legendre transform.

Theorem 6.16 ($C^{2,\alpha}$ regularity up to Γ). *Let $u \in P_1(M)$ be a solution of Problem* **A** *or* **B**, *such that*

$$\Gamma(u) = \{x_n = f(x')\} \cap B_1, \quad \Omega(u) = \{x_n > f(x')\} \cap B_1,$$

with $f \in C^{1,\alpha}(B_1')$. Then u is $C^{2,\alpha}$ on $B_\delta \cap [\Omega(u) \cup \Gamma(u)]$ with $\delta > 0$ and the norm depending only on n, M, and $\|f\|_{C^{1,\alpha}(B_1')}$.

Proof. Consider the transformation of variables $y = T(x)$, $x \in B_1$, given by

$$y_i = x_i, \quad i = 1, \ldots, n-1, \quad y_n = x_n - f(x').$$

This transformations flattens the free boundary, in the sense that it maps $\Omega(u)$ to $\{y_n > 0\}$ and $\Gamma(u)$ to $\{y_n = 0\}$. Moreover we will have that

$$T(\Omega(u)) \supset B_\rho^+, \quad T(\Gamma(u)) \supset B_\rho'$$

for some $\rho > 0$ depending only on the Lipschitz norm of f. Also note that the transformation T is readily invertible with $x = T^{-1}(y)$ given by

$$x_i = y_i, \quad i = 1, \ldots, n-1, \quad x_n = y_n + f(y').$$

Next, for a given direction $e \in \mathbb{R}^n$ consider the function

$$v(y) = v_e(y) = \partial_e u(T^{-1}(y)).$$

Since the transformation T^{-1} is $C^{1,\alpha}$ and $\partial_e u \in C^{0,1}(\Omega(u) \cup \Gamma(u))$, we also have $v \in C^{0,1}(B_\rho^+ \cup B_\rho')$. Moreover, since $\partial_e u$ is harmonic in $\Omega(u)$ and vanishes on $\Gamma(u)$, we see that v is a weak solution of the elliptic equation in the divergence form

$$\begin{aligned} \operatorname{div}(A(y)\nabla_y v) = 0 \quad &\text{in } B_\rho^+, \\ v = 0 \quad &\text{on } B'_\rho, \end{aligned}$$

where the coefficient matrix $A(y) \in C^\alpha(B_\rho^+ \cup B'_\rho)$. Besides, $A(y)$ will be uniformly elliptic there with the ellipticity depending on the Lipschitz norm of f. Then, applying the regularity up to the boundary of the solutions of the divergence form equations with Hölder coefficients (see Gilbarg-Trudinger [**GT01**, Corollary 8.36]), we obtain that $v \in C^{1,\alpha}(B_{\rho/2}^+ \cup B'_{\rho/2})$, with the norm now depending on the $C^{1,\alpha}$-norm of f. This immediately implies that $\partial_e u \in C^{1,\alpha}(B_\delta \cap [\Omega \cup \Gamma])$ for some $\delta > 0$, again depending on f. Since e was arbitrary, we obtain that $D^2 u \in C^\alpha(B_\delta \cap [\Omega \cup \Gamma])$. □

6.4.2. Partial hodograph-Legendre transformation. Here we establish the key result for the partial hodograph-Legendre transformation, which is immediately applicable in our Problems **A** and **B**.

Theorem 6.17. *Let Ω be a bounded open set in $\mathbb{R}^n$, Γ a relatively open subset of $\partial\Omega$, and the function u satisfy*

$$\Delta u = f \quad \text{in } \Omega, \qquad u = |\nabla u| = 0 \quad \text{on } \Gamma,$$

in the classical sense. Assume additionally that

- *Γ is C^1 (i.e. locally a graph of a C^1 function);*
- *$u \in C^2(\Omega \cup \Gamma)$;*
- *$f \in C^{k,\alpha}(\Omega \cup \Gamma)$ for some $k = 1, 2, \ldots$, and $0 < a \leq f \leq b < \infty$ in Ω.*

Then Γ is $C^{k+1,\alpha}$ regular and $u \in C^{k+1,\alpha}(\Omega \cup \Gamma)$, with the norms on $K \Subset \Omega \cup \Gamma$ depending only on K, $\Omega \cup \Gamma$, a, b, n, $\|f\|_{C^{k,\alpha}(\Omega\cup\Gamma)}$, $\|u\|_{C^2(\Omega\cup\Gamma)}$, as well as the modulus of continuity of $D^2 u$.

Furthermore, if f is real-analytic, then so are Γ and u in $\Omega \cup \Gamma$.

Proof. Suppose that $0 \in \Gamma$ and that $e_n = (0, 0, \ldots, 1)$ is the inward unit normal of $\partial\Omega$ at 0. Then we have

$$\partial_{x_i x_j} u(0) = 0 \quad \text{for } i = 1, \ldots, n, \ j = 1, \ldots, n-1.$$

This implies that

$$\partial_{x_n x_n} u(0) = f(0) > 0.$$

Now consider the so-called partial hodograph transformation $y = T(x)$ given by

$$y_n = -u_{x_n}, \quad y_j = x_j, \quad j = 1, \ldots, n-1. \tag{6.10}$$

Since $\det(D_x y) = -\partial_{x_n x_n} u < 0$ at the origin, there is a small $r > 0$, depending on the modulus of continuity of $D^2 u$, such that T is invertible in $\overline{\Omega} \cap B_{2r}$ and maps it onto an open neighborhood of the origin in $\{y_n \leq 0\}$, while $\Gamma \cap B_{2r}$ is mapped onto a relatively open subset of $\{y_n = 0\}$. Further, define

$$v(y) = x_n y_n + u(x), \tag{6.11}$$

known as the partial Legendre transform of u. Direct computations show that

$$\begin{aligned} v_{y_n} &= x_n, \quad v_{y_j} = u_{x_j}, \quad j = 1, \dots, n-1, \\ u_{x_n x_n} &= -\frac{1}{v_{y_n y_n}}, \quad u_{x_n x_j} = \frac{v_{y_n y_j}}{v_{y_n y_n}}, \quad j = 1, \dots, n-1, \\ u_{x_j x_k} &= u_{y_j y_k} - \frac{v_{y_n y_j} v_{y_n y_k}}{v_{y_n y_n}}, \quad j, k = 1, \dots, n-1. \end{aligned}$$

Thus, v satisfies

$$L(D_y^2 v) := \sum_{i=1}^{n-1} \partial_{y_i y_i} v - \frac{1}{\partial_{y_n y_n} v} - \frac{1}{\partial_{y_n y_n} v} \sum_{i=1}^{n-1} (\partial_{y_i y_n} v)^2 = f(T^{-1}(y)).$$

We claim that the fully nonlinear operator $L(D_y^2 v)$ is uniformly elliptic on $T(\Omega \cap B_{2r})$, for small $r > 0$. Treating L as a function on the set of symmetric $n \times n$ matrices $M = \{m_{ij}\}_{i,j=1}^n$, it is enough to verify that the matrix $\{\partial_{m_{ij}} L\}_{i,j=1}^n$ is positive definite at $M = D_y^2 v(0)$. The straightforward computation shows that

$$\begin{aligned} \partial_{m_{jj}} L &= 1, \quad j = 1, \dots, n-1, \\ \partial_{m_{nn}} L &= \frac{1}{m_{nn}^2} \left(1 + \sum_{j=1}^{n-1} m_{jn}^2 \right), \\ \partial_{m_{jk}} L &= 0, \quad j, k = 1, \dots, n-1, \quad j \neq k \\ \partial_{m_{jn}} L &= -2 \frac{m_{jn}}{m_{nn}}, \quad j = 1, \dots, n-1, \end{aligned}$$

making $\partial_{m_{ij}} L$ a diagonal matrix with elements $1, \dots, 1, 1/f(0)^2$ at $M = D_y^2 v(0)$, which is obviously positive definite. This proves that $L(D_y^2 v)$ is uniformly elliptic in $T(\Omega \cap B_{2r})$ for a small $r > 0$.

Further, note that the right-hand side $f(T^{-1}(y))$ can be written as $f(y', v_{y_n})$, and therefore v satisfies

$$\tilde{L}(D_y^2 v, D_y v, y) = L(D_y^2 v) - f(y', v_{y_n}) = 0$$

in $T(\Omega \cap B_{2r})$. Moreover, v vanishes on $T(\Gamma \cap B_{2r})$, which lies in $\{y_n = 0\}$. Since $\tilde{L}$ is uniformly elliptic and $C^{k,\alpha}$ regular in all its variables, we can apply Agmon-Douglis-Nirenberg [**ADN59**, Theorem 11.1] to obtain that

$v \in C^{k+2,\alpha}$ in $T(\overline{\Omega} \cap B_r)$. Considering the inverse transformation $x = T^{-1}(y)$ given by

$$y \mapsto x = (y_1, \ldots, y_{n-1}, \partial_{y_n} v),$$

we find that both u and $\Gamma \cap B_r$ are $C^{k+1,\alpha}$ regular.

If f is real-analytic, then $\tilde{L}$ is real-analytic in all its variables and applying the results in Morrey [**Mor08**, §6.7], we obtain that v is real-analytic and consequently so are Γ and u. This completes the proof of the theorem. □

As an immediate corollary from Theorems 6.16 and 6.17, we obtain the real analyticity of the free boundary in Problems **A** and **B** near regular points.

Theorem 6.18 (Real analyticity of the free boundary). *Let $u \in P_1(M)$ be a solution of Problem* **A** *or* **B** *such that*

$$\Gamma(u) = \{x_n = f(x')\} \cap B_1, \quad \Omega(u) = \{x_n > f(x')\} \cap B_1,$$

with $f \in C^{1,\alpha}(B'_1)$. Then $f \in C^\infty(B'_\delta)$, with $\delta > 0$ and the norms $\|f\|_{C^k(B'_\delta)}$ depending only n, M, and $\|f\|_{C^{1,\alpha}(B'_1)}$. Moreover, f is real-analytic on B'_δ.

Notes

The idea of proving the $C^{1,\alpha}$ regularity from the Lipschitz regularity of the free boundary, based on the boundary Harnack principle, originated in the short note of Athanasopoulos-Caffarelli [**AC85**].

The partial hodograph-Legendre transformation in free boundary problems was developed by Isakov [**Isa75, Isa76**] and independently by Kinderlehrer-Nirenberg [**KN77, KN78a, KN78b, KN78c**]. The method was also generalized to include systems in Kinderlehrer-Nirenberg-Spruck [**KNS78, KNS79**]. It must be mentioned that the minimal regularity assumption required by the method of hodograph-Legendre transformation was provided by Caffarelli [**Caf77**], thus completing the proof of the higher regularity of the free boundary in the obstacle problem.

The C^1 regularity of the free boundaries near branch points in Problem **C** was first proved in Shahgholian-Weiss [**SW06**] in dimension 2 via an Aleksandrov reflection approach and then in Shahgholian-Uraltseva-Weiss [**SUW07**] in arbitrary space dimension. The counterexample of $C^{1,\mathrm{Dini}}$ regularity of the free boundary near branch points in §6.3.2 is due to Shahgholian-Uraltseva-Weiss [**SUW07**]; however, a similar example in the context of the touch of the free and fixed boundaries was already known from Uraltseva [**Ura96**].

The parabolic counterparts of many results in this chapter can be found in Caffarelli-Petrosyan-Shahgholian [**CPS04**] for Problem **A** and Shahgholian-Uraltseva-Weiss [**SUW09**] for Problem **C**.

Exercises

6.1. Give an alternative proof of part (iii) in Lemma 6.1 by compactness argument. Namely, suppose that (iii) fails for $C_0 = j$ and a solution u_j and pass to a limit as $j \to 0$ to arrive at a contradiction.

6.2. Complete the proof of Theorem 6.2.

6.3. A bounded domain $\Omega \subset \mathbb{R}^n$ is called *nontangentially accessible* (NTA) if there exist M, $r_0 > 0$ such that

(a) Ω satisfies the *corkscrew condition*: for any $x^0 \in \partial\Omega$ and $0 < r \leq r_0$ there exists $a_r(x^0) \in \Omega$ such that
$$M^{-1}r \leq \operatorname{dist}(a_r(x^0), \partial\Omega), \quad |a_r(x^0) - x^0| \leq r;$$

(b) $\mathbb{R}^n \setminus \overline{\Omega}$ satisfies the corkscrew condition;

(c) Ω satisfies the *Harnack chain condition*: for any $x^1, x^2 \in \Omega$, with $\operatorname{dist}(x^i, \Omega) \geq \varepsilon$ and $|x^1 - x^2| \leq A\varepsilon$, there exists a chain of $N = N(A, M)$ nontangential balls $B^1, \ldots, B^N$ in Ω, joining x^1 and x^2, i.e. $x^1 \in B^1$, $x^2 \in B^N$, $B^j \cap B^{j+1} \neq \emptyset$, $j = 1, \ldots, N-1$, and
$$M^{-1}\operatorname{diam}(B^j) \leq \operatorname{dist}(B^j, \partial\Omega) \leq M \operatorname{diam}(B^j).$$

Show that bounded Lipschitz domains are NTA.

Hint: First show that the half-ball B_1^+ is NTA. Then note that the property of being NTA is invariant under bi-Lipschitz transformations. Use that to show that the domains of the type $\Omega = \{x_n > f(x')\} \cap B_1$ with Lipschitz f are NTA.

6.4. Prove the existence of the solution u_M of the classical obstacle problem (6.9) for any $M \in [0, 1]$ and show that it is $C^{1,\alpha}$ on $\overline{Q}$, uniformly in M. Moreover, prove that the mapping $M \mapsto u_M$ is continuous from $[0, 1]$ to $C^{1,\alpha}(\overline{Q})$.

Hint: Consider even reflections with respect to the appropriate edges of Q.

Chapter 7

The singular set

In this chapter we study the singular set in Problem **A** and by this essentially complete the study of the free boundary for Problem **A**. This also completes the study of the free boundary in Problem **C**, since near singular one-phase points we can apply the results for Problem **A**.

In §7.1 we give a general characterization of singular points in energetic and geometric terms, and in §7.2 we show that the solutions can be well approximated near singular points by the polynomial solutions. We then discuss explicit examples of Schaeffer of solutions that exhibit analytic cusps (§7.3). Then, in §7.4 we use a monotonicity formula of Monneau to show the uniqueness and continuous dependence of blowups, which leads to a structural theorem for the singular set. In Problem **A**, we do not get the uniqueness of blowups; however, we can still show the existence and continuous dependence of "tangent" spaces, by using careful estimates based on the ACF monotonicity formula (§7.5). This gives the same structural theorem for the singular set as in the classical obstacle problem.

7.1. The characterization of the singular set

Recall that for a solution u of Problem **A**, a free boundary point x^0 is called singular if the rescalings

$$u_r(x) = u_{x^0,r}(x) = \frac{u(x^0 + rx)}{r^2}$$

converge over a subsequence $r = r_j \to 0$ to a homogeneous quadratic polynomial

$$q(x) = \frac{1}{2}(x \cdot Ax),$$

for a symmetric $n \times n$ matrix A with $\operatorname{Tr} A = 1$. We then know (by Theorem 3.23) that all blowups of u at x^0 are polynomial, with possibly different matrices A for different subsequences $r_j \to 0$. In what follows, we will denote by $\Sigma(u)$ the set of all singular points of u.

In what follows, we give two more characterizations of singular points: the first, energetic, based on the balanced energy

$$\omega(x^0) = W(0+, u, x^0),$$

and the second, geometric, based on the thickness function

$$\delta(r, u, x^0) = \frac{1}{r} \min \operatorname{diam}(\Omega^c(u) \cap B_r(x^0)).$$

Proposition 7.1. *Let u be a solution of Problem* **A** *in D. Then for $x^0 \in \Gamma(u)$ the following statements are equivalent:*

(i) $x^0 \in \Sigma(u)$;

(ii) $\omega(x^0) = \alpha_n$;

(iii) $\lim_{r\to 0} \delta(r, u, x^0) = 0$.

Proof. The equivalence of (i) and (ii) is contained in Proposition 3.32.

The implication (i) $\Rightarrow$ (iii) is essentially contained in the proof of Theorem 6.2: if for given $\sigma > 0$ one has $\delta(r, u, x^0) > \sigma$ for a small enough r, then $\partial_e u_{x^0,r} \geq 0$ in $B_{c_n\sigma}$ for a cone of directions $e \in \mathcal{C}_{c_n\sigma}$, implying that any blowup u_0 will have the same property and therefore cannot be polynomial. This means that $\delta(r, u, x^0) \to 0$ as $r \to 0$.

Finally, to show that (iii) $\Rightarrow$ (i), consider a blowup $u_0 = \lim_{j\to\infty} u_{x^0,r_j}$ for some $r_j \to 0$. Since $\Delta u_{x^0,r_j} = 1$ in B_1 except a strip of width $\delta(r_j, u, x^0) \to 0$, we will obtain that

$$\Delta u_0 = 1 \quad \text{a.e. in } B_1,$$

and therefore u_0 must be polynomial. This implies that $x^0 \in \Sigma(u)$. $\square$

In fact, Theorem 6.3 gives a more precise geometric information about singular points.

Lemma 7.2. *Let $u \in P_1(M)$ be a solution of Problem* **A** *and $0 \in \Sigma(u)$. Then there exists a modulus of continuity $\sigma(r)$, depending only on n and M, such that*

$$\delta(r, u) \leq \sigma(r) \quad \textit{for any} \quad 0 < r < 1.$$

7.2. Polynomial solutions

Consider the class

$$\mathcal{Q} = \{q(x) \text{ homogeneous quadratic polynomial} : \Delta q = 1\},$$
$$\mathcal{Q}^+ = \{q \in \mathcal{Q} : q \geq 0\}.$$

Essentially, the elements of $\mathcal{Q}$ ($\mathcal{Q}^+$) are the blowups of solutions u of Problem **A** (the classical obstacle problem) at a singular points x^0. Clearly, polynomials $q \in \mathcal{Q}$ are themselves solutions.

Each member of $\mathcal{Q}$ is uniquely represented as

$$q(x) = \frac{1}{2}(x \cdot Ax),$$

where A is a symmetric $n \times n$ matrix with $\operatorname{Tr} A = 1$. In fact, $A = D^2 q$. Then note that the free boundary

$$\Gamma(q) = \ker A$$

consists completely of singular points. The dimension $d = \dim \ker A$ can be any integer from 0 to $n-1$. More specifically, if after a rotation of coordinate axes we represent

$$q(x) = \sum_{i=1}^{m} \lambda_i x_i^2 \quad \text{with } \lambda_i > 0,\ 1 \leq m \leq n,$$

where $\lambda_1, \ldots, \lambda_m$ are nonzero eigenvalues of the matrix A, then $d = n - m$.

The case $d = n-1$ deserves a special attention. For definiteness, consider $q_0(x) = \frac{1}{2}x_n^2$. Then the free boundary is a hyperplane $\Pi = \{x_n = 0\}$ of codimension one, which is as smooth as it can be. Nevertheless, all points on Π are singular, since $\Omega^c(q_0) = \Pi$ is "thin". In contrast, for the half-space solution $h_0(x) = \frac{1}{2}(x_n^+)^2$, the free boundary is still Π; however, this time it consists of regular points, since $\Omega(h_0) = \{x_n \leq 0\}$ is "thick". This means that the singular set can be as large as the set of regular points.

The next result, in a sense, is complementary to Lemma 5.13.

Lemma 7.3 (Approximation by polynomial solutions). *Let $u \in P_1(M)$ be a solution of Problem* **A** *with $0 \in \Sigma(u)$. Then there exists a modulus of continuity $\sigma(r)$ depending only on M and n and such that for any $0 < r < 1$ there exists a homogeneous quadratic polynomial $q^r \in \mathcal{Q}$ such that*

$$\|u - q^r\|_{L^\infty(B_{2r})} \leq \sigma(r) r^2,$$
$$\|\nabla u - \nabla q^r\|_{L^\infty(B_{2r})} \leq \sigma(r) r.$$

Moreover, if $u \geq 0$ (i.e. u solves the classical obstacle problem), then one can take $q^r \in \mathcal{Q}^+$.

Proof. Equivalently, we can show that for any $\varepsilon > 0$ there exists $r_\varepsilon > 0$ depending only on M and n and such that for any $0 < r \le r_\varepsilon$ we can find $q^r \in \mathcal{Q}$ that satisfies

$$|u - q^r| \le \varepsilon r^2, \quad |\nabla u - \nabla q^r| \le \varepsilon r \quad \text{in } B_{2r}.$$

Arguing by contradiction, assume that this statement fails for some $\varepsilon > 0$. Then we can find sequences $r_j \to 0$ and $u_j \in P_1(M)$ with $0 \in \Sigma(u_j)$ and such that for any $q \in \mathcal{Q}$ either one or the other of the following holds:

$$\|u_j - q\|_{L^\infty(B_{2r_j})} \ge \varepsilon r_j^2, \quad \|\nabla u_j - \nabla q\|_{L^\infty(B_{2r_j})} \ge \varepsilon r_j.$$

Consider now the sequence

$$v_j(x) = (u_j)_{r_j}(x) = \frac{u_j(r_j x)}{r_j^2}, \quad |x| < 1/r_j.$$

Clearly, $v_j \in P_{1/r_j}(M)$ and therefore, over a subsequence, $v_j \to v_0$ in $C^{1,\alpha}_{\mathrm{loc}}(\mathbb{R}^n)$. We claim that $v_0 \in \mathcal{Q}$. Indeed, by Lemma 7.2,

$$\delta(\rho, v_j) = \delta(\rho r_j, u_j) \le \sigma(\rho r_j) \to 0,$$

for any $\rho > 0$. This means that v_j satisfies $\Delta v_j = 1$ in B_ρ, except a strip of width $\sigma(\rho r_j) \to 0$. Therefore v_0 satisfies $\Delta v_0 = 1$ a.e. in $\mathbb{R}^n$, and consequently, by Liouville theorem, $v_0 \in \mathcal{Q}$. On the other hand, from the assumptions on u_j with $q = v_0$ we must have either one or the other of the following:

$$\|v_j - v_0\|_{L^\infty(B_2)} \ge \varepsilon, \quad \|\nabla v_j - \nabla v_0\|_{L^\infty(B_2)} \ge \varepsilon,$$

which clearly contradicts the convergence $v_j \to v_0$ in $C^{1,\alpha}_{\mathrm{loc}}(\mathbb{R}^n)$.

For the second part of the lemma, the proof for $u \ge 0$ works exactly the same way by replacing $\mathcal{Q}$ with $\mathcal{Q}^+$. □

7.3. Examples of singularities

Here we describe examples of analytic singularities, constructed by Schaeffer [**Sch77**], that may occur in the classical obstacle problem. Heuristically, the existence of such singularities can be seen as follows. If Λ is a set bounded by an analytic curve with a possible singularity, then by the Cauchy-Kovalevskaya theorem we may find a function that solves $\Delta u = 1$ in a neighborhood outside Λ and satisfies $u = \partial_\nu u = 0$ on $\partial\Lambda$. Shrinking Λ one may then make $u > 0$ in a neighborhood of Λ. More precise construction is given below. We will use both real and complex notation in what follows.

Let D be a simply connected domain in $\mathbb{C} = \mathbb{R}^2$ and Λ a closed subset of D with a piecewise C^1 boundary. Let $\Omega = D \setminus \Lambda$ and suppose that we are given a conformal mapping $\phi : R_{1,2} \to \Omega$ continuous up to the boundary. Here by $R_{a,b}$ we denote the ring $\{a < |z| < b\}$ for $0 < a < b$. We will assume

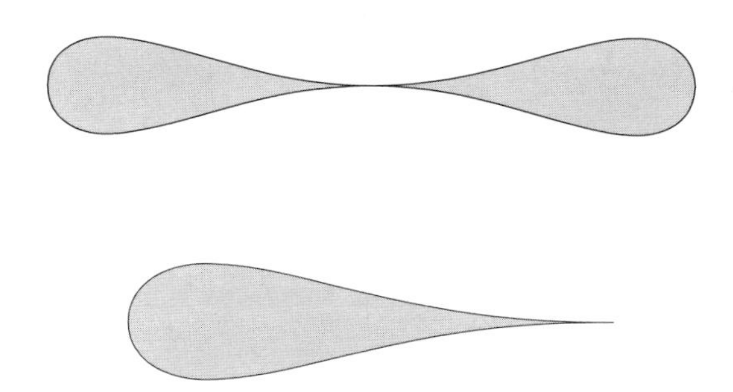

Figure 7.1. The shaded regions correspond to the coincidence set Λ in Schaeffer's examples with ϕ_1 (top) and ϕ_2 (bottom) respectively.

that ϕ maps ∂B_2 to ∂D and ∂B_1 to $\partial\Lambda$. Additionally, we will assume that D and Λ are symmetric with respect to the real axis and that

$$\phi(\overline{z}) = \overline{\phi(z)}, \quad z \in \overline{R_{1,2}}.$$

If now ϕ is holomorphically extensible to a mapping $R_{1/2,2} \to \mathbb{C}$, we can define

$$f(z) = -\phi(1/\phi^{-1}(z)), \quad z \in \Omega.$$

Note that $f(z)$ defined as above will be continuous up to $\partial\Lambda$ if we put

$$f(z) = -\overline{z}, \quad z \in \partial\Lambda.$$

Next consider

$$v(z) = \operatorname{Re} \int f(z)dz, \quad z \in \Omega,$$

the real part of the indefinite integral of f. Even though $\int f(z)dz$ can be multivalued due to Ω not being simply connected, we claim that v is well defined. Indeed, note that

$$\operatorname{Re} \int_{\partial\Lambda} f(z)dz = \int_{\partial\Lambda} \operatorname{Re}(\overline{z}dz) = \int_{\partial\Lambda} d(|z|^2/2) = 0$$

and therefore $\operatorname{Re}\int_\gamma f(z)dz = 0$ for any closed curve γ in Ω. Consequently v is well defined in Ω. Of course v is harmonic in Ω and

$$\partial_x v - i\partial_y v = f(z), \quad z \in \Omega.$$

Therefore $\nabla v(z) = -(x, y)$ continuously for $z = x + iy \in \partial\Lambda$. By choosing the constant of integration, we can arrange also that $v(z) = -|z|^2/2$ on $\partial\Lambda$. Define now

$$u(z) = \begin{cases} \frac{1}{4}|z|^2 + \frac{1}{2}v(z), & z \in \Omega, \\ 0, & z \in \Lambda. \end{cases}$$

Then $u \in C^1(D) \cap C^2(D \setminus \partial\Lambda)$ and it is easy to see that

$$\Delta u = \chi_\Omega \quad \text{in} \quad D, \qquad u = |\nabla u| = 0 \quad \text{on} \quad \Lambda,$$

where the first equality is in the sense of distributions.

More specifically, consider two mappings given by

$$\begin{aligned} \phi(z) = \phi_i(z) &= (z + 1/z)/2 + \varepsilon P_i(z)(z - 1/z)/2, \quad i = 1, 2,\\ P_1(z) &= z^2 + 2 + 1/z^2,\\ P_2(z) &= (z - 2 + 1/z)^2, \end{aligned}$$

for $\varepsilon > 0$. They have the properties discussed above for ε sufficiently small. The corresponding boundaries of Λ have the parametrizations

$$\begin{aligned} \phi_1(e^{i\theta}) &= \cos\theta + i4\varepsilon\cos^2\theta\sin\theta,\\ \phi_2(e^{i\theta}) &= \cos\theta + i32\varepsilon\cos(\theta/2)\sin^5(\theta/2) \end{aligned}$$

and have the shapes similar to those in Fig. 7.1 with singularities at $z_0 = 0$ (self-touching) and $z_0 = 1$ (cusp).

It can be shown that the solutions constructed above are also nonnegative if $\varepsilon > 0$ is sufficiently small, thus making u a solution of the classical obstacle problem with a coincidence set Λ. The proof is outlined in Exercise 7.2.

7.4. Singular set: classical obstacle problem

To study the set of singular points in the case of the classical obstacle problem we take the approach of Monneau [**Mon03**], based on a monotonicity formula, specifically tailored for singular points.

7.4.1. Monneau's monotonicity formula.

Theorem 7.4 (Monneau's monotonicity formula). *Let $u \in P_R(M)$ be a solution of the classical obstacle problem with $0 \in \Sigma(u)$. Then for any $q \in \mathcal{Q}^+$, the functional*

$$r \mapsto M(r, u, q) = \frac{1}{r^{n+3}} \int_{\partial B_r} (u - q)^2 dH^{n-1}$$

is monotone nondecreasing for $r \in (0, 1)$.

Remark 7.5. Before giving the proof, note that we have the following nice rescaling property:

$$M(r, u, q) = M(1, u_r, q),$$

and that we can actually introduce Monneau's functional centered at any $x^0 \in \Sigma(u)$ by

$$M(r, u, q, x^0) = M(1, u_{x^0,r}, q) = \int_{\partial B_1} (u_{x^0,r} - q)^2$$
$$= \frac{1}{r^{n+3}} \int_{\partial B_r(x^0)} [u(x) - q(x - x^0)]^2.$$

The theorem above then trivially generalizes for $M(r, u, q, x^0)$.

Proof. We first compute the derivative of M. Let $w = u - q$; then

$$\begin{aligned}\frac{d}{dr} M(r, u, q) &= \frac{d}{dr}\left(\frac{1}{r^{n+3}} \int_{\partial B_r} w^2(x)\right)\\ &= \frac{d}{dr} \int_{\partial B_1} \frac{w^2(ry)}{r^4}\\ &= \int_{\partial B_1} \frac{2w(ry)(ry \cdot \nabla w(ry) - 2w(ry))}{r^5}\\ &= \frac{2}{r^{n+4}} \int_{\partial B_r} w(x \cdot \nabla w - 2w).\end{aligned}$$

From the assumptions on u and q we have $W(0+, u) = W(r, q) = \alpha_n$, where W is Weiss's energy functional. Therefore, letting $w = u - q$, we can write

$$\begin{aligned}&W(r, u) - W(0+, u) = W(r, u) - W(r, q)\\ &= \frac{1}{r^{n+2}} \int_{B_r} (|\nabla w|^2 + 2\nabla w \cdot \nabla q + 2w) - \frac{2}{r^{n+3}} \int_{\partial B_r} (w^2 + 2wq)\\ &= \frac{1}{r^{n+2}} \int_{B_r} |\nabla w|^2 - \frac{2}{r^{n+3}} \int_{\partial B_r} w^2 + \frac{2}{r^{n+3}} \int_{\partial B_r} w(x \cdot \nabla q - 2q)\\ &= \frac{1}{r^{n+2}} \int_{B_r} |\nabla w|^2 - \frac{2}{r^{n+3}} \int_{\partial B_r} w^2\\ &= \frac{1}{r^{n+2}} \int_{B_r} (-w\Delta w) + \frac{1}{r^{n+3}} \int_{\partial B_r} w(x \cdot \nabla w - 2w).\end{aligned}$$

On the other hand, we have

$$w\Delta w = (u - q)(\Delta u - \Delta q) = \begin{cases} 0 & \text{on } \{u > 0\},\\ q \geq 0 & \text{a.e. on } \{u = 0\}.\end{cases}$$

Thus, combining the computations above, we arrive at

$$\frac{d}{dr} M(r, u, q) \geq \frac{2}{r}[W(r, u) - W(r, 0+)] \geq 0. \qquad \square$$

As an immediate corollary we obtain the following.

Corollary 7.6. *Let u be a solution of the classical obstacle problem and $x^0 \in \Sigma(u)$. Then the blowup of u at x^0 is unique.*

Proof. Let $q_0 = \lim_{r_j \to 0} u_{x^0,r_j}$ be a blowup over a certain sequence $r_j \to 0$. Since $q_0 \in \mathcal{Q}^+$, we may apply Theorem 7.4 with $q = q_0$. Since the convergence $u_{x^0,r_j} \to q_0$ is in $C^{1,\alpha}_{\text{loc}}$ in $\mathbb{R}^n$, we have, in particular, that

$$M(r_j, u, q_0, x^0) = \int_{\partial B_1} (u_{x^0,r_j} - q_0)^2 \to 0.$$

Now observe that the monotone function $r \mapsto M(r, u, u_0, x^0)$ converges to 0 over a sequence $r_j \to 0$ iff it converges to 0 for all $r \to 0$. Thus, we obtain that

$$\int_{\partial B_1} (u_{x^0,r} - q_0)^2 \to 0 \quad \text{as } r \to 0,$$

and therefore every convergent subsequence of $u_{x^0,r}$ must converge to q_0. □

Next, we show that not only the blowup is unique at singular points but that we can also control the convergence rate in a uniform fashion.

Proposition 7.7 (Continuous dependence of blowups). *Let u be a solution of the classical obstacle problem in an open set D in $\mathbb{R}^n$. For $x^0 \in \Sigma(u)$ denote by q_{x^0} the blowup of u at x^0. Then for any $K \Subset D$, there exists a modulus of continuity $\sigma(r)$ depending only on n, $\|u\|_{L^\infty(D)}$,* dist$(K, \partial D)$, *such that*

$$|u(x) - q_{x^0}(x - x^0)| \le \sigma(|x - x^0|)|x - x^0|^2,$$
$$|\nabla u(x) - \nabla q_{x^0}(x - x^0)| \le \sigma(|x - x^0|)|x - x^0|,$$

provided $x^0 \in \Sigma(u) \cap K$, $x \in D$. Besides, the mapping $x^0 \mapsto q_{x^0}$ is continuous from $\Sigma(u)$ to $\mathcal{Q}^+$ with

$$\|q_{x^1} - q_{x^2}\|_{L^2(\partial B_1)} \le \sigma(|x^1 - x^2|),$$

provided $x^1, x^2 \in \Sigma(u) \cap K$.

Remark 7.8. Note that since $\mathcal{Q}^+$ is contained in the finite-dimensional linear space of homogeneous quadratic polynomials, any two norms on this space are equivalent, and therefore we can replace the $L^2(\partial B_1)$-norm in the last estimate say by the $L^\infty(B_1)$-norm or even $C^1(B_1)$-norm.

Proof. Without loss of generality assume that $D = B_1$ and $K = \overline{B_{1/2}}$. Assume also that $x^0 = 0$. For $r > 0$ let $q^r \in \mathcal{Q}^+$ be the approximating polynomial as in Lemma 7.3, and for $\varepsilon > 0$ let $r_\varepsilon > 0$ be such that $\sigma(r) \le \varepsilon$

in the same lemma. Then, applying Monneau's monotonicity formula for u with $q = q^{r_\varepsilon}$, we will have

$$\int_{\partial B_1} (u_r - q^{r_\varepsilon})^2 \le \int_{\partial B_1} (u_{r_\varepsilon} - q^{r_\varepsilon})^2 \le C_n\varepsilon^2, \quad 0 < r \le r_\varepsilon.$$

Letting $r \to 0+$, we will obtain

$$\int_{\partial B_1} (q_0 - q^{r_\varepsilon})^2 \le C_n\varepsilon^2.$$

Since both q_0 and q^{r_ε} are polynomials from $\mathfrak{Q}^+$ (see Remark 7.8), we also have

$$|q_0 - q^{r_\varepsilon}| \le C_n\varepsilon, \quad |\nabla q_0 - \nabla q^{r_\varepsilon}| \le C_n\varepsilon \quad \text{in } B_1.$$

Therefore, combining this with Lemma 7.3, we obtain

$$|u - q_0| \le C_n\varepsilon r_\varepsilon^2, \quad |\nabla u - \nabla q_0| \le C_n\varepsilon r_\varepsilon \quad \text{in } B_{r_\varepsilon}.$$

Considering the mapping $r \mapsto \sigma(r)$, which is the inverse of $\varepsilon \mapsto r_\varepsilon$, we obtain the first part of the proposition.

To prove the second part of the proposition, we may assume without loss of generality that $x^2 = 0$. Now, if $x^1 \in B_{r_\varepsilon/2} \cap \Sigma(u)$, then as above, we have

$$\int_{\partial B_1} (q_{x^1} - q^{r_\varepsilon})^2 \le \int_{\partial B_1} (u_{x^1, r_\varepsilon/2} - q^{r_\varepsilon})^2 \le C_n\varepsilon^2.$$

Hence

$$\int_{\partial B_1} (q_{x^1} - q_0)^2 \le C_n\varepsilon^2 \quad \text{if } x^1 \in B_{r_\varepsilon/2} \cap \Sigma(u).$$

This completes the proof of the theorem. □

7.4.2. Structure of the singular set. For a singular point $x^0 \in \Sigma(u)$, let $q_{x^0} = \frac{1}{2}(x \cdot A_{x^0}x)$ be the blowup of u at x^0. Because of the uniqueness of the blowup, it is natural to think of

$$\Pi_{x^0} = \Gamma(q_{x^0}) = \ker A_{x^0}$$

as the "tangent" of $\Sigma(u)$ at x^0, and define the "dimension" of $\Sigma(u)$ at x^0 by

$$d_{x^0} = \dim \Pi_{x^0}.$$

Note that since A_{x^0} is never zero, $d_{x^0} \in \{0, 1, \dots, n-1\}$.

Theorem 7.9 (Structure of the singular set). *Let u be a solution of the obstacle problem in an open set D in $\mathbb{R}^n$. Then for any $d \in \{0, 1, \dots, n-1\}$, the set*

$$\Sigma^d(u) = \{x^0 \in \Sigma(u) : d_{x^0} = d\}$$

is contained in a countable union of d-dimensional C^1 manifolds.

The proof is based on two tools from real analysis: Whitney's extension theorem [**Whi34**] and the implicit function theorem.

Lemma 7.10 (Whitney's extension theorem). *Let E be a compact set in $\mathbb{R}^n$ and $f : E \to \mathbb{R}$ a certain mapping. Suppose that for any $x^0 \in E$ there exists a polynomial p_{x^0} of degree m such that*

(i) *$p_{x^0}(x^0) = f(x^0)$ for any $x^0 \in E$;*

(ii) *$|D^k p_{x^0}(x^1) - D^k p_{x^1}(x^1)| = o(|x^0 - x^1|^{m-k})$ for any $x^0, x^1 \in E$ and $k = 0, \dots, m$.*

Then f extends to a C^m function on $\mathbb{R}^n$ such that

$$f(x) = p_{x^0}(x) + o(|x - x^0|^m),$$

for all $x^0 \in E$. □

Proof of Theorem 7.9. Let K be a compact subset of D and $E = \Sigma(u) \cap K$. Note that since $\Sigma(u)$ is closed, E is also compact.

We next claim that the family of polynomials $p_{x^0}(x) = q_{x^0}(x - x^0)$, $x^0 \in E$, satisfies the assumptions of Whitney's extension theorem on E with $f = 0$ and $m = 2$. Indeed, it is clear that condition (i) is satisfied: $p_{x^0}(x^0) = q_{x^0}(0) = 0$. Thus, we need to verify the compatibility conditions in (ii) for $k = 0, 1, 2$.

1) $k = 0$. We need to show that

$$|q_{x^0}(x^1 - x^0)| = o(|x^1 - x^0|^2), \quad \text{for } x^0, x^1 \in \Sigma(u) \cap K.$$

This is easily verified from the estimate

$$|u(x) - q_{x^0}(x - x^0)| \le \sigma(|x - x^0|)|x - x_0|^2$$

in Proposition 7.7, by plugging $x = x^1 \in \Sigma(u) \cap K$ and noticing that $u(x^1) = 0$.

2) $k = 1$. Then the compatibility condition is

$$|\nabla q_{x^0}(x^1 - x^0)| = o(|x^1 - x^0|), \quad \text{for } x^0, x^1 \in \Sigma(u) \cap K.$$

This time it is verified from the second estimate

$$|\nabla u(x) - \nabla q_{x^0}(x - x^0)| \le \sigma(|x - x^0|)|x - x_0|$$

in Proposition 7.7, by plugging $x = x^1 \in \Sigma(u) \cap K$ and noticing that $\nabla u(x^1) = 0$.

3) $k = 2$. In terms of the coefficient matrix of q_{x^0}, we need to show that

$$A_{x^0} - A_{x^1} = o(1).$$

But this is equivalent to the continuity of the mapping $x^0 \mapsto q_{x^0}$, which is established again in Proposition 7.7.

Now let f be a C^2 function on $\mathbb{R}^n$, provided by Whitney's extension theorem. To complete the proof, observe that

$$\Sigma(u) \cap K \subset \{\nabla f = 0\} = \bigcap_{i=1}^{n} \{\partial_{x_i} f = 0\}.$$

For $x^0 \in \Sigma^d(u) \cap K$, we can arrange the coordinate axes so that the vectors $e_1, \ldots, e_{n-d}$ are the eigenvalues of $D^2 f(x^0) = A_{x^0}$. We then have that

$$\det D^2_{(x_1,\ldots,x_{n-d})} f(x^0) \neq 0.$$

Since f is C^2, the implicit function theorem implies that

$$\bigcap_{i=1}^{n-d} \{\partial_{x_i} f = 0\}$$

is a d-dimensional C^1 manifold in a neighborhood of x^0. This completes the proof of the theorem. □

7.5. Singular set: Problem A

In this section we prove an analogue of Theorem 7.9 for solutions of Problem **A**. One complication that we encounter is that we no longer have the uniqueness of blowups (or of the limits of approximation polynomials q^r as $r \to 0$), but instead we have the uniqueness of the limits of $|\nabla q^r|^2$. As a consequence, we will need to use the ACF monotonicity formula instead of that of Monneau. Then a further delicate analysis shows the existence and continuous dependence of "tangent" spaces at singular points, which ultimately leads to the structural theorem for the singular set.

In the next lemmas, for a direction $e \in \mathbb{R}^n$, we use the notation

$$\phi_e(r, v) = \Phi(r, (\partial_e v)^+, (\partial_e v)^-),$$

as before, where Φ is the ACF functional.

Lemma 7.11. *Let $u \in P_1(M)$ be a solution of Problem* **A** *with $0 \in \Sigma(u)$ and q^r, $0 < r < 1/2$, and the modulus of continuity σ as in Lemma 7.3. Then for any direction e,*

$$|\phi_e(r, u) - \phi_e(r, q^r)| \leq C\sigma(r)^{1/3},$$

where $C = C(M, n)$.

Proof. We want to estimate

$$I(r, (\partial_e u)^+) = I(1, (\partial_e u_r)^+) = \int_{\{\partial_e u_r > 0\} \cap B_1} \frac{|\nabla \partial_e u_r|^2 dx}{|x|^{n-2}}$$

in terms of $I(r,(\partial_e q^r)^+) = I(1,(\partial_e q^r)^+)$. First, by Lemma 7.3 we have

$$\|\nabla u_r - \nabla q^r\|_{L^\infty(B_1)} \le \sigma(r). \tag{7.1}$$

Further, we can also estimate the second derivatives if we stay away from the free boundary. Note that

$$\Omega^c(u_r) \cap B_1 \subset S_{\sigma(r)},$$

where S_h is a strip of width h (in a certain direction that we do not specify). Then outside $S_{\sigma(r)^{1/3}}$ we apply the interior $C^{1,1}$ estimates to obtain

$$\begin{aligned} \|D^2 u_r - D^2 q^r\|_{L^\infty(B_1 \setminus S_{\sigma(r)^{1/3}})} &\le \frac{C_n \|u_r - q^r\|_{L^\infty(B_2)}}{\sigma(r)^{2/3}} \\ &\le C_n \sigma(r)^{1/3}. \end{aligned} \tag{7.2}$$

To proceed, let A^r be the Hessian of q^r so that $q^r(x) = \frac{1}{2}(x \cdot A^r x)$. Then by $C^{1,1}$ estimates for u^r, combined with (7.2), we immediately have $|A^r| \le C(M,n)$. Now let

$$U^r = \{\partial_e u_r > 0\} \cap B_1, \quad V^r = \{\partial_e q^r = x \cdot A^r e > 0\} \cap B_1,$$

and

$$U^r_\pm = \{\partial_e q^r = x \cdot A^r e > \pm\sigma(r)\} \cap B_1.$$

Then by (7.1)

$$U^r_+ \subset U^r \subset U^r_-,$$

and therefore

$$\int_{U^r_+} \frac{|\nabla \partial_e u_r|^2 dx}{|x|^{n-2}} \le \int_{U^r} \frac{|\nabla \partial_e u_r|^2 dx}{|x|^{n-2}} \le \int_{U^r_-} \frac{|\nabla \partial_e u_r|^2 dx}{|x|^{n-2}}. \tag{7.3}$$

Next, using (7.2), we estimate

$$\begin{aligned} &\left| \int_{U^r_\pm} \frac{|\nabla \partial_e u_r|^2 dx}{|x|^{n-2}} - \int_{U^r_\pm} \frac{|\nabla \partial_e q^r|^2 dx}{|x|^{n-2}} \right| \\ &\quad \le C \int_{S_{\sigma(r)^{1/3}} \cap B_1} \frac{dx}{|x|^{n-2}} + C\sigma(r)^{1/3} \int_{B_1 \setminus S_{\sigma(r)^{1/3}}} \frac{dx}{|x|^{n-2}} \\ &\quad \le C\sigma(r)^{1/3}, \end{aligned} \tag{7.4}$$

where $C = C(M,n)$. Furthermore,

$$\begin{aligned} &\left| \int_{U^r_\pm} \frac{|\nabla \partial_e q^r|^2 dx}{|x|^{n-2}} - \int_{V^r} \frac{|\nabla \partial_e q^r|^2 dx}{|x|^{n-2}} \right| \\ &\qquad \le C \frac{\sigma(r)}{|A^r e|} |A^r e|^2 \le C\sigma(r). \end{aligned} \tag{7.5}$$

Note that in both (7.4) and (7.5) we have used that

$$\int_{S_h\cap B_1}\frac{dx}{|x|^{n-2}}\le\int_{-h}^{h}\int_{\{|x'|\le1\}}\frac{dx'}{|x'|^{n-2}}dx_n\le Ch,\quad 0<h<1.$$

Now, putting (7.3)–(7.5) together, we obtain

$$|I(1,(\partial_e u_r)^+)-I(1,(\partial_e q^r)^+)|\le C\sigma(r)^{1/3}.$$

We have an analogous estimate for $I(1,(\partial_e u_r)^-)$, and taking the product we obtain

$$|\phi_e(1,u_r)-\phi_e(1,q^r)|\le C\sigma(r)^{1/3},$$

which implies the statement of the lemma. □

Lemma 7.12. *Let u and q^r be as Lemma 7.11 and $q^r(x)=\frac{1}{2}(x\cdot A^r x)$ for a symmetric matrix A^r with $\operatorname{Tr} A^r=1$. Then for $0<r_1\le r_2<1/2$ and any unit vector e,*

$$|A^{r_1}e|^2\le|A^{r_2}e|^2+C\sigma(r_2)^{1/3}.$$

Proof. Recall that $\phi_e(r,u)$ is monotone nondecreasing in r. Then the estimate follows from Lemma 7.11 and the observation that

$$\phi_e(r,q^r)=C_n|A^r e|^2.$$ □

Lemma 7.13. *Let A^r be as in Lemma 7.12. Then there exists a unique matrix B_0 such that if $A^r\to A$ for a subsequence $r=r_k\to 0$, then*

$$A^2=B_0.$$

In particular, $\Pi_0=\ker A=\ker B_0$ is also unique.

Proof. Let A' and A'' be the limits of A^r as $r=r_k'\to0$ and $r=r_k''\to0$ respectively. Then by Lemma 7.12

$$|A'e|^2=|A''e|^2,$$

for any unit vector e. On the other hand, since A is a symmetric matrix, $Ae\cdot Ae=e\cdot A^2e$ and therefore the above equality implies that $A'^2=A''^2$. □

Now let u be a solution of Problem **A** in an open set D in $\mathbb{R}^n$ and $x^0\in\Sigma(u)$. If we denote by $q^r_{x^0}\in\mathcal{Q}$ the approximating polynomials of $u(\cdot-x^0)$ as in Lemma 7.3 and by $A^r_{x^0}$ their Hessians, then by Lemma 7.13 we obtain the unique limit

$$B_{x^0}=\lim_{r\to0+}(A^r_{x^0})^2.$$

Thus,

$$\Pi_{x^0}=\ker B_{x^0}$$

can be viewed as the "tangent" space of $\Sigma(u)$ at x^0. Therefore, similarly to the classical obstacle case, we also define

$$d_{x^0} = \dim \Pi_{x^0}.$$

Note again that B_{x^0} is never zero and therefore $d_{x^0} \in \{0, 1, \dots, n-1\}$.

We can now state the main result of this section.

Theorem 7.14 (Structure of the singular set: Problem **A**). *Let u be a solution of Problem **A** in an open set D in $\mathbb{R}^n$ and let Π_{x^0}, d_{x^0} for $x^0 \in \Sigma(u)$ be defined as above. For $d \in \{0, 1, \dots, n-1\}$ define*

$$\Sigma^d(u) = \{x \in \Sigma(u) : d_{x^0} = d\}.$$

Then $\Sigma^d(u)$ is contained in a countable union of d-dimensional C^1 manifolds.

The proof is again based on Whitney's extension theorem and the implicit function theorem. However, it is much more delicate than in the nonnegative case.

For $x^0 \in \Sigma^d(u)$, let $A = \lim_{j\to\infty} A_{x^0}^{r_j}$ for some sequence $r_j \to 0$. Also let $q(x) = \lim_{j\to\infty} q_{x^0}^{r_j}(x) = \frac{1}{2}(x \cdot Ax)$. We may choose a system of coordinates such that

$$q(x) = \frac{1}{2}\sum_{i=1}^{n-d} \lambda_i x_i^2, \quad \text{with } |\lambda_i| > 0,$$

where $\lambda_1, \dots, \lambda_{n-d}$ are the nonzero eigenvalues of A and $\lambda_1^2, \dots, \lambda_{n-d}^2$ are those of $B_0 = A^2$. For a given $a > 0$ we then define

$$\Sigma_a^d(u) = \{x^0 \in \Sigma^d(u) : |\lambda_i| \geq a,\ i = 1, \dots, n-d\}.$$

Lemma 7.15. *Let $u \in P_1(M)$ be a solution of Problem **A** with $0 \in \Sigma_a^d(u)$ and $\Pi_0 = \ker B_0$ as in Lemma 7.13. For a given $\varepsilon > 0$ define the cones*

$$\mathcal{C}_\varepsilon(\Pi_0) = \{\operatorname{dist}(x, \Pi_0) \geq \varepsilon |x|\}.$$

Then for any direction $e \in \mathcal{C}_\varepsilon(\Pi_0)$ we have

$$\phi_e(0+, u) \geq C_n a^4 \varepsilon^4.$$

Proof. Let $q(x) = \lim_{j\to\infty} q^{r_j}(x) = \frac{1}{2}(x \cdot Ax)$ for a certain sequence $r_j \to 0$. Note that $\Pi_0 = \ker A$. Now, because of Lemma 7.11, we will be done once we show that

$$\phi_e(1, q) \geq C_n a^4 \varepsilon^4.$$

The latter inequality is straightforward. Indeed, for $e \in \mathcal{C}_\varepsilon(\Pi_0)$ we have

$$|\nabla \partial_e q|^2 = |Ae|^2 \geq a^2 \operatorname{dist}(e, \Pi_0)^2 \geq a^2 \varepsilon^2,$$

which readily implies the claim. □

Lemma 7.16. *Let $u \in P_1(M)$ be a solution of Problem* **A** *with $0 \in \Sigma_a^d(u)$ and $\Pi_0 = \ker B_0$ as in Lemma 7.13. Then for any $\varepsilon > 0$ there exists $r_\varepsilon > 0$, depending only on ε, n, M, d, and a, and such that*

$$\mathcal{C}_\varepsilon(\Pi_0) \cap B_{r_\varepsilon} \subset \Omega(u).$$

Proof. If the assertion of the lemma fails, then we can find sequences $u_j \in P_1(M)$ with $0 \in \Sigma_a^d(u_j)$ and $x^j \in \Omega^c(u_j) \cap \mathcal{C}_\varepsilon(\Pi_0(u_j))$ with $r_j = |x^j| \to 0$. Consider the rescalings

$$v_j(x) = (u_j)_{r_j}(x) = \frac{u_j(r_j x)}{r_j^2}.$$

Then $v_j \in P_{1/r_j}(M)$, and over a subsequence we may assume that $v_j \to v_0$ in $C^{1,\alpha}_{\mathrm{loc}}(\mathbb{R}^n)$. It is easy to see that v_0 is a polynomial solution. Indeed, by Lemma 7.2, for any $\rho > 0$,

$$\delta(\rho, v_j) = \delta(\rho r_j, u_j) \leq \sigma(\rho \rho_j) \to 0.$$

This means that v_j satisfies $\Delta v_j = 1$ in B_ρ, except in a strip of width $\sigma(\rho r_j) \to 0$. Therefore v_0 satisfies $\Delta v_0 = 1$ a.e. in $\mathbb{R}^n$, and consequently, by Liouville's theorem, $v_0 \in \mathcal{Q}$.

Further, passing to a subsequence if necessary, we may assume that $e^j = x^j / r_j \to e^0 \in \partial B_1$. By our construction $e^j \in \Omega^c(v_j)$ and therefore $e^0 \in \Omega^c(v_0)$. Since $v_0 \in \mathcal{Q}$, we immediately conclude that $\partial_{e_0} v_0 \equiv 0$ in $\mathbb{R}^n$. On the other hand, by Lemma 7.15 we have

$$\phi_{e^j}(1, v_j) = \phi_{e^j}(r_j, u_j) \geq \phi_{e^j}(0+, u_j) \geq C_n \alpha^4 \varepsilon^4.$$

Passing to the limit (justified by Proposition 3.17(v)) we obtain

$$\phi_{e^0}(1, v_0) \geq C_n a^4 \varepsilon^4,$$

which clearly contradicts the fact that $\partial_{e^0} v_0 \equiv 0$. $\square$

To state the next result, for $x^0 \in \Sigma(u)$ let π_{x^0} be the orthogonal projection to the orthogonal complement of the subspace $\Pi_{x^0} = \ker B_{x^0}$. In particular

$$\ker \pi_{x^0} = \Pi_{x^0}.$$

Besides, if $x^0 \in \Sigma_a^d(u)$, then

$$a^2 |\pi_{x^0} e|^2 \leq e \cdot B_{x^0} e \tag{7.6}$$

for any unit vector e in $\mathbb{R}^n$.

Proposition 7.17. *Let u be a solution of Problem* **A** *in an open set D in $\mathbb{R}^n$. Then for any $K \Subset D$, $d \in \{0, 1, \ldots, n-1\}$, and $a > 0$, there* *exists a modulus of continuity* $\sigma(r)$, *depending only on* n, d, a, $\|D^2u\|_{L^\infty(D)}$, $\operatorname{dist}(K, \partial D)$, *and such that*

$$\|\pi_{x^1} - \pi_{x^0}\|_{L^\infty(B_1)} \le \sigma(|x^1 - x^0|), \tag{7.7}$$

$$|\pi_{x^0}(x^1 - x^0)| \le \sigma(|x^1 - x^0|)|x^1 - x^0|, \tag{7.8}$$

provided $x^0, x^1 \in \Sigma_a^d(u) \cap K$.

Proof. Without loss of generality we may assume that $D = B_1$, $K = \overline{B_{1/2}}$, and $x^0 = 0$.

1) The estimate (7.7) is an immediate corollary of Lemma 7.16. Just observe that

$$\|\pi_{x^1} - \pi_{x^0}\|_{L^\infty(B_1)}$$

measures the distance between the subspaces Π_{x^1} and Π_{x^0}.

2) To show (7.8), let $q^r(x) = \frac{1}{2}(x \cdot A^r x)$ be as in Lemma 7.3. Choosing $r = |x^1|$ and using that $|\nabla u(x^1)| = 0$, we then have

$$|A^r x^1| = |\nabla q^r(x^1)| = |\nabla u(x^1) - \nabla q^r(x^1)| \le \sigma(r) r.$$

On the other hand, writing $x^1 = re$ for a unit vector e, using (7.6) and Lemma 7.12, we have

$$\begin{aligned} a^2|\pi_0(x^1)|^2 = a^2 r^2 |\pi_0 e|^2 &\le r^2(e \cdot B_0 e) \\ &\le r^2|A^r e|^2 + C\sigma(r) r^2 \\ &= |A^r x^1|^2 + C\sigma(r) r^2 \le \sigma(|x^1|)|x^1|^2. \end{aligned}$$

(In the above computation σ stands for a generic modulus of continuity.) This implies (7.8). □

We can now give the proof of Theorem 7.14.

Proof of Theorem 7.14. For given $d \in \{0, \dots, n-1\}$, $a > 0$, and compact $K \Subset D$ consider the set $E = \Sigma_a^d(u) \cap K$. Note that at this point we do not know if E is closed. Consider then the family of polynomials

$$p_{x^0}(x) = (x - x^0)\pi_{x^0}(x - x^0).$$

Then Proposition 7.17 implies that

$$\begin{aligned} |p_{x^0}(x^1)| &\le \sigma(|x^0 - x^1|)|x^0 - x^1|^2, \\ |\nabla p_{x^0}(x^1)| &\le \sigma(|x^0 - x^1|)|x^0 - x^1|, \\ \|D^2 p_{x^0} - D^2 p_{x^1}\| &\le \sigma(|x^0 - x^1|), \end{aligned}$$

for $x^0, x^1 \in E$ that satisfy conditions (ii) in Whitney's extension theorem (Lemma 7.10). Because of the uniformity of these estimates, we may extend the family $\{p_{x^0}\}$ to be defined for $x^0 \in \overline{E}$, while keeping the same estimates.

We then can apply Whitney's extension theorem and the implicit function theorem to conclude the proof as for Theorem 7.9. □

Notes

Schaeffer [**Sch77**] was the first to construct examples of solutions of the obstacle problem with strictly superharmonic real-analytic obstacles for which the free boundary exhibits a singularity. The exposition in §7.3 mainly follows this paper.

The uniqueness of blowups at singular points in the classical obstacle problem was first established by Caffarelli-Riviére [**CR77**] in dimension 2. A similar result in higher dimensions had to wait until the paper of Caffarelli [**Caf98a**] and was based on the application of the ACF monotonicity formula. Subsequently, a similar result was obtained by Weiss [**Wei99b**], with the use of what we now call Weiss's monotonicity formula. The simplest proof, however, was given by Monneau [**Mon03**], where he introduced a new monotonicity formula, tailor-made for singular points. Our treatment of the classical obstacle problem follows that of Monneau.

In Problem **A**, the uniqueness of blowups at singular points is not known; however, the singular set has a uniquely defined tangent space (see §7.5). In Problem **A**, only the method of Caffarelli [**Caf98a**] adapted in Caffarelli-Shahgholian [**CS04**] is known to work.

The structure of the singular set in the classical obstacle problem as in §7.4.2 was discovered by Caffarelli [**Caf98a**]. In the case of Problem **A**, it is due to Caffarelli-Shahgholian [**CS04**]. It also noteworthy that for the classical obstacle problem in dimension 2, Monneau [**Mon03**] was able to show the discreteness of the set of singular points that are on the boundary of the interior of the coincidence set.

In Problem **B**, very little is known about the singular set; all the existing methods seem to fail.

Exercises

7.1. Prove the following geometric characterization of singular points, in addition to Proposition 7.1. For a solution u of Problem **A**, $x^0 \in \Gamma(u)$ is singular iff the lower Lebesgue density of $\Omega^c(u)$ at x^0 is zero, i.e.

$$\liminf_{r\to 0} \frac{|\Omega^c(u) \cap B_r(x^0)|}{|B_r|} = 0.$$

7.2. The goal of this exercise it to show that in Schaeffer's construction in §7.3 with ϕ being one of the functions ϕ_i, $i = 1, 2$, one has that $u > 0$ in Ω, provided the parameter $\varepsilon > 0$ is sufficiently small. We follow the proof in [**Sch77**].

First note that applying the inverse mapping ϕ^{-1}, we can write the condition $u > 0$ in Ω as

$$\begin{aligned} U(z) &= |\phi(z)|^2/2 - \operatorname{Re}\int \phi(1/z)\phi'(z)dz \\ &= \operatorname{Re}\int [\overline{\phi(z)} - \phi(1/z)]\phi'(z)dz > 0 \quad \text{in } R_{1,2}. \end{aligned}$$

(a) Using a representation

$$\phi(z) = \phi_+(z) + \varepsilon\phi_-(z), \quad \text{with } \phi_\pm(1/z) = \pm\phi_\pm(z)$$

show that

$$U(z) = a(z) + \varepsilon b(z) + \varepsilon^2 c(z),$$

where

$$a(z) = [\operatorname{Im}\phi_+(z)]^2,$$

and hence

$$a(re^{i\theta}) = \frac{1}{4}(r - r^{-1})^2 \sin^2\theta,$$

implying that $a > 0$ on $\partial B_2 \setminus \{\pm 2\}$.

(b) Show that

$$b(x) = 2\int_1^x \phi_-(t)\phi_+(t)dt, \quad \text{for } x \in [0, 2],$$

and consequently $b(2) > 0$. Similarly show that $b(-2) > 0$ and conclude that $U > 0$ on ∂B_2, provided $\varepsilon > 0$ is small. Note also that $U = 0$ on ∂B_1.

(c) Show that U satisfies the minimum (maximum) principle in $R_{1,2}$, since $|\nabla U|$ does not vanish there. Combining with the result of part (b), conclude that $U > 0$ in $R_{1,2}$.
To this end, use the representation

$$(\partial_x - i\partial_y)U = [\overline{\phi(z)} - \phi(1/z)]\phi'(z)$$

and observe that $\phi'(z) \neq 0$ on $R_{1,2}$, since ϕ is one-to-one there, and that $\overline{\phi(z)} = \phi(1/z)$ iff

$$\operatorname{Im}\phi_+(z) = 0 \quad \text{and} \quad \operatorname{Re}\phi_-(z) = 0,$$

which does not happen at any point of the ring $R_{1,2}$.

7.3. Let the matrices A^{r_1} and A^{r_2} be as in Lemma 7.12 and additionally assume that they are nonnegative. Show that

$$|A^{r_1} - A^{r_2}| \leq C\sigma(r_2)^{1/3}.$$

This gives an alternative proof that the polynomials q^r have a unique limit as $r \to 0$ for solutions of the classical obstacle problem.

Hint: Let $B = A^{r_1} - A^{r_2}$. Then $\operatorname{Tr} B = 0$ and let $\lambda = \lambda_{\max} \geq 0$ be the largest eigenvalue of B and e the corresponding unit eigenvector. We have

$$|A^{r_1}e|^2 = |Be + A^{r_2}e|^2 \leq |A^{r_2}e|^2 + C\sigma(r_2)^{1/3},$$

and since $Be = \lambda e$, this gives

$$\lambda^2 + 2\lambda(e \cdot A^{r_2}e) \leq C\sigma(r_2)^{1/3}.$$

Using the nonnegativity of A^{r_2} find an estimate for $\lambda_{\max}$, and from the condition $\operatorname{Tr} B = 0$ extend the estimate to $\lambda_{\min}$, the minimum eigenvalue of B.

7.4. Let u be a solution of Problem **A** and $x^0 \in \Sigma(u)$. Show that even though we do not know the uniqueness of blowups at x^0, the limit

$$\lim_{r \to 0} |\nabla u_{x^0,r}|^2$$

does exist. Here $u_{x^0,r}(x) = r^{-2}u(x^0 + rx)$.

Hint: Use Lemmas 7.3 and 7.13.

7.5. Let u be a solution of Problem **A** and $x^0 \in \Sigma(u)$. Show that if all nonzero eigenvalues of B_{x^0} are simple (i.e. their eigenspaces are one-dimensional), then there exists $\lim_{r\to 0} A^r_{x^0}$, and consequently the blowups of u at x^0 are unique.

Chapter 8

Touch with the fixed boundary

In this chapter we study the behavior of solutions of obstacle-type problems near so-called fixed boundaries. Namely, given a solution u in an open set D with Dirichlet boundary values g on ∂D we want to describe the behavior of the free boundary $\Gamma(u)$ near ∂D. Because of the technical difficulty of the problem, we will restrict ourselves to the case of flat boundaries and zero Dirichlet data.

In §8.1 we define the contact points and the appropriate classes of local and global solutions. We next classify all global solutions, as well as the possible blowups at contact points (§8.2). This allows us to establish the tangential behaviour of the free boundary near contact points, as well as its C^1 character for solutions of Problems **A** and **C** (§8.3). Finally, in §8.4, we show the uniqueness of blowups and the existence of nontangential limits for second derivatives.

For convenience, we also recall the notation that will be frequently used in this chapter:

$$\mathbb{R}^n_+ := \mathbb{R}^{n-1} \times (0,\infty), \quad \Pi := \mathbb{R}^{n-1} \times \{0\},$$
$$B^+_R := B_R \cap \mathbb{R}^n_+, \qquad B'_R := B_R \cap \Pi.$$

8.1. Contact points

For solutions u in an open set D, recall that the free boundary was defined by $\Gamma(u) = \partial\Omega(u) \cap D$, where $\Omega(u)$ was specified for Problems **A**, **B**, and **C**. We now extend the notion of the free boundary by adding the so-called contact points with ∂D (see Fig. 8.1).

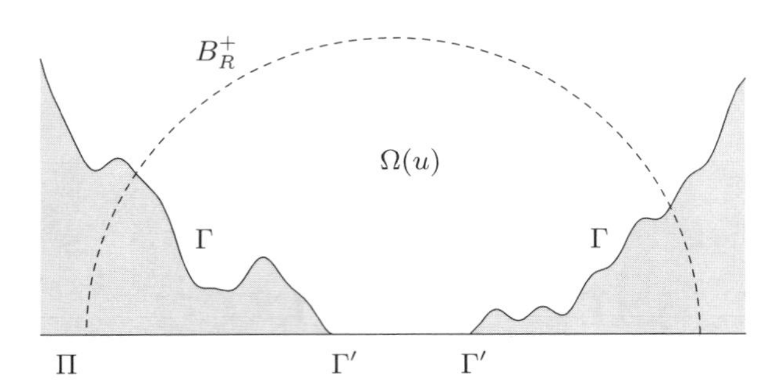

Figure 8.1. Contact points

Definition 8.1 (Contact points)**.** We say that x^0 is a *contact point* of the free boundary for a solution u of Problem **A**, **B**, or **C** in D if

$$x^0 \in \partial\Omega(u) \cap \partial D \quad \text{and} \quad |\nabla u(x^0)| = 0.$$

We will denote the set of all contact points by $\Gamma'(u)$.

Because of the $C^{1,1}$ regularity up to the boundary (see Theorem 2.17), we may restrict our study to the following classes of local solutions.

Definition 8.2 (Local solutions near flat boundaries)**.** For given $R, M > 0$ let $P_R^+(M)$ be the class of solutions u of Problem **A**, **B**, or **C**, such that $u \in C^{1,1}(B_R^+ \cup B'_R)$ and

- $\|D^2 u\|_{L^\infty(B_R^+)} \leq M$,
- $u = 0$ on B'_R,
- $\Gamma(u) \cup \Gamma'(u) \neq \emptyset$ ($\Gamma^0(u) \cup \Gamma'(u) \neq \emptyset$ in Problem **C**).

If $x^0 \in \Gamma(u) \cup \Gamma'(u)$, then we say that $u \in P_R^+(x^0, M)$. Note that this notation is somewhat different from the one used in the interior case. Here x^0 is not the center of the ball but a point on $\Gamma(u) \cup \Gamma'(u)$ and we have

$$P_R^+(M) = \bigcup_{x^0 \in B_R^+ \cup B'_R} P_R^+(x^0, M).$$

Definition 8.3 (Global solutions in half-spaces)**.** Formally taking $R = \infty$ in Definition 8.2, we obtain the class of global solutions $P_\infty^+(M)$ (respectively $P_\infty^+(x^0, M)$).

As in the interior case, we will use the blowup approach in our analysis of the boundary behavior. Thus, we will be considering limits of the rescalings of the types

$$(u_j)_{x^j, r_j}(x) = \frac{u_j(x^j + r_j x) - u(x^j)}{r_j^2}, \quad u_j \in P_R^+(x^j, M).$$

To exclude the possibility of vanishing of such limits, it is important to have nondegeneracy, similarly to the interior case. We state the relevant nondegeneracy statements in Exercises 8.4 and 8.5. One particular corollary from the nondegeneracy is the following proposition.

Proposition 8.4. *Let $u_j \in P_R^+(0, M)$ and $u_j \to u_0$ in $C^{1,\alpha}_{\mathrm{loc}}(B_R \cup B'_R)$. Then $u_0 \in P_R^+(0, M)$.*

Proof. The $C^{1,\alpha}$ convergence up to the boundary implies that $|\nabla u_0(0)| = 0$. Thus, the only fact that needs to be verified is that $0 \in \partial\Omega(u_0)$. By nondegeneracy (see Exercises 8.4 and 8.5) for any $\rho \in (0, R)$ we have that

$$\sup_{B_\rho^+} u_j \geq \frac{\rho^2}{2n} \qquad \text{in Problems } \mathbf{A}, \mathbf{B},$$

$$\sup_{B_\rho^+} |u_j| \geq c_0(n, \lambda_\pm)\rho^2 \quad \text{in Problem } \mathbf{C}.$$

Then similar estimates hold for u_0, implying that 0 is indeed a contact point. □

8.2. Global solutions in half-spaces

The first step towards the local regularity of the free boundary, as in the interior case, is the classification of global solutions. In the next theorem we show that the global solutions of Problems **A**, **B**, and **C** are at most two-dimensional and give their complete description.

Theorem 8.5. *Let $u \in P_\infty^+(M)$ be a global solution of Problem* **A**, **B**, *or* **C**. *Then either $\Gamma(u)$ or $\Gamma'(u)$ is empty. More precisely, after a possible rotation of coordinate system in Π, we have one of the following representations for the function u.*

– *In Problem* **A**:

$$u(x) = \frac{x_n^2}{2} + a x_n x_1 + \alpha x_n, \qquad a, \alpha \in \mathbb{R},$$

$$u(x) = \frac{[(x_n - b)^+]^2}{2}, \qquad b > 0.$$

– *In Problem* **B**:

$$u(x) = \frac{x_n^2}{2} + a x_n x_1 + \alpha x_n, \qquad a, \alpha \in \mathbb{R},$$

$$u(x) = \frac{[(x_n - a)^-]^2}{2} + \frac{[(x_n - b)^+]^2}{2} - \frac{a^2}{2}, \qquad 0 \leq a \leq b \leq \infty.$$

– *In Problem* **C**:

$$u(x) = \pm\lambda_\pm \frac{[(x_n - b)^+]^2}{2}, \quad b \geq 0.$$

Corollary 8.6. *In Theorem 8.5, assume additionally that the origin is a contact point, i.e. $u \in P_\infty^+(0, M)$. Then, after a possible rotation of coordinate system in Π, we have*

$$u(x) = \frac{x_n^2}{2} + a x_n x_1, \quad a \in \mathbb{R}, \qquad \textit{in Problems } \mathbf{A}, \mathbf{B},$$

$$u(x) = \pm\lambda_\pm \frac{(x_n^+)^2}{2} \qquad \textit{in Problem } \mathbf{C}.$$

Proof of Theorem 8.5. 1) We start by showing that u depends at most on two variables. Let $u \in P_\infty^+(M)$ and consider a shrinkdown u_∞ of u at the origin,

$$u_\infty(x) = \lim_{R_k \to \infty} u_{R_k}(x), \quad u_R(x) = \frac{u(Rx)}{R^2}.$$

Note that the rescalings u_R are defined in $\mathbb{R}^n_+ \cup \Pi$ and $u_R\big|_\Pi = 0$. As usual, we have that

$$u_{R_k} \to u_\infty \quad \text{in } W^{2,p}_{\text{loc}}(\mathbb{R}^n_+ \cup \Pi), \quad u_\infty\big|_\Pi = 0.$$

Fix $e \perp e_n$ and extend u, $\partial_e u$, and $\partial_e u_\infty$ by zero to $\mathbb{R}^n_-$. Then we have

$$\begin{aligned} \phi_e(r, u_\infty) &= \lim_{R_k \to \infty} \phi_e(r, u_{R_k}) \\ &= \lim_{R_k \to \infty} \phi_e(R_k r, u) = \phi_e(\infty, u) =: c_e, \end{aligned}$$

that is, $\phi_e(r, u_\infty)$ does not depend on r. Since $\partial_e u_\infty \equiv 0$ on $\mathbb{R}^n_-$, we conclude that $c_e = 0$ (see Theorem 2.9). Moreover,

$$0 \leq \phi_e(r, u) \leq \phi_e(\infty, u) = c_e = 0.$$

Thus, we obtain from this that neither $\partial_e u$ nor $\partial_e u_\infty$ changes its sign for any $e \perp e_n$. This implies that $u = u(x_1, x_n)$ and $u_\infty = u_\infty(x_1, x_n)$ after a suitable rotation of coordinate axes in Π (see Exercise 8.3).

2) Now we prove that either u is independent of x_1, so that $u = u(x_n)$, or

$$\text{Int}\, \Omega^c(u_\infty) = \emptyset. \tag{8.1}$$

For this purpose we extend $v = \partial_{x_1} u$ by odd reflection to $\mathbb{R}^n_-$, still denoting the extension by v. Without loss of generality we will also assume that $v \geq 0$ in $\mathbb{R}^n_+$ (otherwise we will flip the direction of e_1). It is easy to see (Exercise 8.8) that

$$\Delta v^\pm \geq 0 \quad \text{in } \mathbb{R}^n.$$

Consider now the rescalings

$$v_R(x) = \frac{v(Rx)}{R}.$$

Since $v \geq 0$ in $\mathbb{R}^n_+$, then

$$\begin{aligned} v_R &\geq 0 \quad \text{in } \mathbb{R}^n_+, \\ v_R &\leq 0 \quad \text{in } \mathbb{R}^n_-, \end{aligned}$$

and (for a subsequence $R_k \to \infty$)

$$v_{R_k} \to v_\infty(x) = \begin{cases} \partial_{x_1} u_\infty(x_1, x_n), & x \in \mathbb{R}^n_+, \\ -\partial_{x_1} u_\infty(x_1, -x_n), & x \in \mathbb{R}^n_-. \end{cases}$$

We have

$$\begin{aligned} \Phi(r, v^\pm) \leq c := \Phi(\infty, v^\pm) &= \lim_{R_k \to \infty} \Phi(R_k r, v^\pm) \\ &= \lim_{R_k \to \infty} \Phi(r, v^\pm_{R_k}) = \Phi(r, v^\pm_\infty), \end{aligned} \tag{8.2}$$

where the right-hand side is strictly positive as long as $\partial_{x_1} u_\infty$ does not vanish identically in $\mathbb{R}^n_+$. From the case of equality in the ACF monotonicity formula (Theorem 2.9), if u_∞ does depend on x_1, then $v_\infty \neq 0$ a.e. in $\mathbb{R}^n_+$, that is, (8.1) holds.

If it happens that $u_\infty = u_\infty(x_n)$, then $\Phi(r, v^\pm_\infty) \equiv 0$ and by (8.2) we obtain that $\Phi(r, v^\pm) \equiv 0$. The latter means that $v \equiv 0$ in $\mathbb{R}^n_+$, that is, $u = u(x_n)$.

3) To finish the proof of the theorem, we consider the cases of Problems **A**, **B** and of Problem **C** separately.

Problems **A**, **B**. If $u = u(x_n)$, we then easily find the possible solutions. Otherwise, from steps 1) and 2) above we know that $u_\infty = u_\infty(x_1, x_n)$ and that $\operatorname{Int} \Omega^c(u_\infty) = \emptyset$. Then the function $v_\infty = \partial_{x_1} u_\infty$ satisfies

$$\Delta v_\infty = 0 \quad \text{a.e. in } \mathbb{R}^n_+, \quad v_\infty\big|_\Pi = 0, \quad |v_\infty(x)| \leq C(|x| + 1) \quad \text{in } \mathbb{R}^n_+,$$

and therefore by the Liouville theorem $v_\infty = a x_n$. Integrating, we conclude that

$$u_\infty(x) = \frac{x_n^2}{2} + a x_1 x_n + \alpha x_n. \tag{8.3}$$

Now, we want to prove a similar representation for u. If $\operatorname{Int} \Omega^c(u) = \emptyset$, then repeating the previous arguments for u_∞ we obtain the desired representation. Therefore, suppose $\operatorname{Int} \Omega^c(u) \neq \emptyset$ and let $B_\rho(z^0) \subset \Omega^c(u)$. Recall that we assume that $\partial_{x_1} u \geq 0$. It follows then that $u \leq u(z^0)$ in the cylinder $(-\infty, z_1^0) \times \{|x_n - z_n^0| < \rho\}$, and since u is subharmonic, from the strong maximum principle we obtain that $u \equiv u(z^0) = c_0$ there, and therefore

$$(-\infty, z_1^0) \times \{|x_n - z_n^0| < \rho\} \subset \Omega^c(u). \tag{8.4}$$

Consider now a sequence

$$u_{(m)}(x) = u(x - me_1), \quad m = 1, 2, \dots .$$

By monotonicity of u in the e_1-direction, for any $x \in \mathbb{R}^n_+$ there exists a finite limit

$$u_{(\infty)}(x_n) = \lim_{m\to\infty} u_{(m)}(x).$$

Indeed, from the inclusion (8.4) we have the uniform estimate $|u_{(m)}(x)-c_0| \leq M|x_n - z^0_n|^2$, for sufficiently large m. Thus, $u_{(\infty)} = u_{(\infty)}(x_n)$ is a one-dimensional solution identically equal to c_0 on $\{|x_n - z^0_n| < \rho\}$. It is easy to see that $u_{(\infty)} \geq c_0$, and since $u_{(\infty)} \leq u_{(m)}(x) \leq u(x)$, we also obtain that $u \geq c_0$ in $\mathbb{R}^n_+$. Hence, we conclude that the shrinkdown $u_\infty \geq 0$. Together with (8.3) this implies that $u_\infty = \frac{1}{2}x_n^2$. Recalling the remark at the end of step 2), we again arrive at the case $u = u(x_n)$ and thereby complete the proof for Problems **A**, **B**.

Problem **C**. We claim that $u_\infty = u_\infty(x_n)$. This again will imply that $u = u(x_n)$ and complete the proof. To prove the claim we use the following two facts:

a) $u_\infty(x) = \lim_{R_k\to\infty} u_{x^0,R_k} = \lim \frac{u(x^0+R_k x)}{R_k^2}$, where $x^0 \in \mathbb{R}^n_+ \cup \Pi$ is an arbitrary point such that $u(x^0) = 0$ (cf. Exercise 5.7).

b) For x^0 as in a) and $r > 0$ consider the Weiss functional $W(r, u, x^0)$ defined by (3.12), where u is extended to $\mathbb{R}^n_-$ as identically 0. For such u we have the differentiation formula

$$\begin{aligned}\frac{d}{dr}W(r,u,x^0) = {} & \frac{2}{r^{n+4}}\int_{\partial B_r(x^0)} |\partial'_{(x^0)}u|^2 dH^{n-1}\\ & + \frac{2x^0_n}{r^{n+2}}\int_{B'_r(x^0)} |\partial_{x_n}u|^2 dH^{n-1};\end{aligned} \tag{8.5}$$

see Exercise 8.1. This implies that $W(r, u, x^0)$ is monotone nondecreasing in r. Since $W(R, u, x^0)$ is uniformly bounded for $R \in [1, \infty)$, there exists a finite limit

$$W(\infty, u, x^0) = \lim_{R_k\to\infty} W(R_k r, u, x^0) = \lim_{R_k\to\infty} W(r, u_{x^0,R_k}) = W(r, u_\infty).$$

Consequently, $dW(r, u_\infty)/dr \equiv 0$. Hence, from the differentiation formula (8.5) it follows that u_∞ is a function homogeneous of degree two and consequently that $\Gamma^*(u_\infty) = \emptyset$ (see the proof of Theorem 3.18). Therefore, the $u^\pm_\infty$ are global solutions of the obstacle problem in $\mathbb{R}^n_+$ that depend only on x_n due to the results for Problem **A**.

This completes the proof of the theorem. □

8.3. Behavior of the free boundary near the fixed boundary

Theorem 8.7. *Let $u \in P_1^+(0,M)$ be a solution of Problem* **A**, **B**, *or* **C**. *There exist a universal constant $r_0 = r_0(n,M) > 0$ and a modulus of continuity $\sigma(r)$ such that*

(i) (Tangential touch) $\Gamma(u) \cap B_{r_0}^+ \subset \{x : x_n \leq \sigma(|x|)|x|\}$.

(ii) (C^1 touch) *In Problems* **A** *and* **C**, *$\partial\Omega(u) \cap B_{r_0}^+$ is the graph of a C^1 function over B'_{r_0}, with normals to $\partial\Omega(u)$ having the modulus of continuity $\sigma(r)$.*

Remark 8.8. The conclusion in part (ii) is not known for Problem **B**. In Problems **A** and **C**, the C^1 regularity cannot be improved beyond $C^{1,\mathrm{Dini}}$. So, in a sense, the result is optimal. Indeed, if we revisit §6.3.2, the function u_{M_0} is a solution of the classical obstacle problem in $B_\delta(A_0) \cap \{x_1 > 0\}$, for some $\delta > 0$, and vanishes on $B_\delta(A_0) \cap \{x_1 = 0\}$. However, $\partial\Omega(u_{M_0})$ is not $C^{1,\mathrm{Dini}}$.

The first part of the theorem is equivalent to the following proposition.

Proposition 8.9. *Let $u \in P_1^+(0,M)$ be a solution of Problem* **A**, **B**, *or* **C**. *Then for any $\varepsilon > 0$, there exists $\rho = \rho_\varepsilon > 0$ such that*

$$\Gamma(u) \cap B_{\rho_\varepsilon}^+ \subset B_{\rho_\varepsilon}^+ \setminus \mathcal{C}_\varepsilon. \tag{8.6}$$

We recall that the cone $\mathcal{C}_\varepsilon$ is defined by

$$\mathcal{C}_\varepsilon := \{x : x_n > \varepsilon|x'|\}.$$

Proof. The proof is by contradiction. Suppose the conclusion of the lemma fails and there exist $u_j \in P_1^+(0,M)$ and $x^j \in \Gamma(u_j) \cap B_1^+$ with $d_j := |x^j| \to 0$ such that $x^j \in \Gamma(u) \cap \mathcal{C}_\varepsilon$. Consider then rescalings

$$\tilde{u}_j(x) = \frac{u_j(d_j x)}{d_j^2}.$$

Now for each function $\tilde{u}_j$ we have a point $\tilde{x}^j = x^j/d_j \in \Gamma(\tilde{u}_j) \cap \partial B_1^+ \cap \mathcal{C}_\varepsilon$. Next, by compactness, over a subsequence, $\tilde{u}_j$ and $\tilde{x}^j$ converge to u_0 and x^0, respectively, with u_0 a global solution and $x^0 \in \overline{\mathcal{C}}_\varepsilon \cap \partial B_1$. Moreover, by Proposition 3.17(iv) and Exercises 8.4 and 8.5 we will have $x^0 \in \Gamma(u_0)$ and $0 \in \Gamma'(u_0)$. Since u_0 is a global solution, this contradicts Theorem 8.5. □

To prove the second part of Theorem 8.7 for Problem **A**, we will need the following approximation lemma.

Lemma 8.10. *Let $u \in P_1^+(M)$ be a solution of Problem* **A**. *For any $\varepsilon > 0$ there exists $r(\varepsilon) = r(\varepsilon, M, n) > 0$ such that if $x^0 \in \Gamma(u) \cap B_{1/2}^+$ and*

$d := x_n^0 < r(\varepsilon)$, then

$$\sup_{B_{2d}^+(x^0)} |u - h| \le \varepsilon d^2, \qquad \sup_{B_{2d}^+(x^0)} |\nabla u - \nabla h| \le \varepsilon d,$$

where

$$h(x) = \frac{1}{2}[(x_n - d)^+]^2.$$

Proof. If the lemma fails for some $\varepsilon > 0$, then there exist $u_j \in P_1^+(M)$ and $x^j \in \Gamma(u_j) \cap B_{1/2}^+$ with $d_j := x_n^j \to 0$ as $j \to \infty$, such that at least one of the following inequalities holds:

$$\sup_{B_{2d_j}^+(x^j)} |u_j - \tfrac{1}{2}[(x_n - d_j)^+]^2| > \varepsilon d_j^2, \qquad \sup_{B_{2d_j}^+(x^j)} |\nabla u_j - (x_n - d_j)^+| > \varepsilon d_j.$$

Set

$$v_j(x) := \frac{u_j(x^j + d_j(x - e_n))}{d_j^2}.$$

Then $v_j \in P_{1/(2d_j)}^+(e_n, M)$ and

$$\left\| v_j - \tfrac{1}{2}[(x_n - 1)^+]^2 \right\|_{C^1(B_2^+(e_n))} > \varepsilon.$$

Next note that since $|\nabla u_j(x^j)| = 0$ and $|D^2 u_j| \le M$, we have $|v_j(x)| \le M|x - e_n|^2$, which makes v_j locally uniformly bounded. Hence, we can choose a subsequence of $\{v_j\}$ converging to $v_0 \in P_\infty^+(e_n, M)$ with the property

$$\left\| v_0 - \tfrac{1}{2}[(x_n - 1)^+]^2 \right\|_{C^1(B_2^+(e_n))} \ge \varepsilon.$$

On the other hand, by Theorem 8.5, $v_0 = \frac{1}{2}[(x_n - 1)^+]^2$, since this is the only global solution in $P_\infty^+(e_n, M)$. Hence, we arrived at a contradiction. □

The next result essentially reduces the study of Problem **C** near contact points to Problem **A**.

Lemma 8.11. *Let $u \in P_1^+(0, M)$ be a solution of Problem* **C**. *Then for some $r_0 = r_0(n, M) > 0$ either $u \ge 0$ or $u \le 0$ in $B_{r_0}^+$. In particular, u is a solution to a one-phase problem in $B_{r_0}^+$.*

Proof. This follows easily by compactness. Indeed, if the lemma fails, then there is a sequence u_j of two-phase solutions such that for each B_{r_j} we have by nondegeneracy (see Exercise 8.5) both of the following estimates:

$$\sup_{B_{r_j}^+} u_j \ge c_0 r_j^2, \qquad -\inf_{B_{r_j}^+} u_j \ge c_0 r_j^2.$$

In particular, any converging subsequence of

$$\tilde{u}_j(x) = \frac{u_j(r_j x)}{r_j^2}$$

will converge to a global solution u_0 with

$$\sup_{B_1^+} u_0 \geq c_0, \qquad -\inf_{B_1^+} u_0 \geq c_0,$$

contradicting the classification of global solution, Theorem 8.5, for Problem **C**. □

We next prove that for solutions of Problem **A**, the free boundary $\partial\Omega$ is a C^1 graph in a narrow strip near Π.

Proposition 8.12. *Let $u \in P_1^+(M)$ be a solution of Problem* **A**. *Then there exists $h = h(n, M) > 0$ such that each connected component of $\partial\Omega(u) \cap B_{1/2} \cap \{0 \leq x_n < h\}$ is a graph $x_n = f(x')$ with $f \in C^1$, $\|f\|_{C^1} \leq C(n, M)$.*

Proof. Let $r(\varepsilon)$ be the constant from Lemma 8.10 and set

$$S_\delta := \{|x'| < 1/2\} \times \{0 < x_n < r(\delta/32n)\}.$$

By Lemmas 8.10 and 4.7, every point $x^0 \in \Gamma(u) \cap S_1$ is regular and $\Gamma(u) \cap B_{\frac{1}{2}x_n^0}(x^0)$ is represented as a graph $x_n = f^{x^0}(x')$ with the Lipschitz constant of f^{x^0} not exceeding 1. In fact, one can prove (see Exercise 8.6) that $u \equiv 0$ in $(x^0 - \mathcal{C}_1) \cap \mathbb{R}^n_+$. The latter means that the entire set $\partial(\Omega^c(u) \cap S_1)$ in the ball $B_{1/2}$ is a graph $x_n = f(x')$ with the Lipschitz constant of f not exceeding 1. By the results of Chapter 6, the function f is actually C^∞ (real-analytic) at the points x' with $0 < f(x') < h := r(1/32n)$.

Next, by Proposition 8.9, the normal vector to $\partial\Omega(u)$ at any contact point $x^0 = (x^{0\prime}, 0)$ exists and coincides with e_n. Moreover, by Lemmas 8.10 and 4.4, the inward normal vector $\nu(y)$ at any $y \in \Gamma(u) \cap S_\delta$ is δ-close to e_n. This implies that f is C^1 at the contact points x^0 and moreover that $|\nabla f(x^{0\prime})| = 0$. To complete the proof of the proposition, we need to show that the normal vectors $\nu(x)$ to $\partial\Omega(u) \cap S_1$ are continuous with the same modulus of continuity for the entire class $P_1^+(M)$.

Indeed, we claim that for any fixed $\delta > 0$ we can find $\rho_\delta = \rho(\delta, n, M) > 0$ such that

$$\text{angle}(\nu(y^1), \nu(y^2)) \leq \delta \quad \text{if } |y^1 - y^2| \leq \rho_\delta, \quad y^1, y^2 \in \Gamma(u) \cap S_1, \tag{8.7}$$

for any $u \in P_1^+(M)$.

It is clear that for any $y^1, y^2 \in \Gamma(u) \cap \overline{S_{\delta/4}}$ the inequality (8.7) holds, since in this case both $\nu(y^1)$ and $\nu(y^2)$ are $(\delta/4)$-close to e_n. If both points, y^1 and y^2, are outside $\overline{S_{\delta/4}}$, then we may use scaling and apply Theorem 6.11. This results in the estimate

$$\rho_\delta \leq \left(\frac{\delta}{2C}\right)^{1/\alpha} r\left(\frac{\delta}{128n}\right),$$

where $C = C(n, M)$ and $\alpha = \alpha(n, M)$ are the constants from Theorem 6.11, and $r(\varepsilon)$ is as in Lemma 8.10. The latter value of ρ_δ also provides the estimate (8.7) in the case when $y^1 \in \overline{S_{\delta/4}}$, $y^2 \notin S_{\delta/4}$. □

Proof of Theorem 8.7. For statement (i) in the theorem, we consider the modulus of continuity $\sigma(t)$ given by the inverse of the mapping $\varepsilon \mapsto \rho_\varepsilon$, for ρ_ε as in Proposition 8.9, and let $r_0 = \rho_{\{\varepsilon=1\}}$.

Statement (ii) for Problem **A** follows from Proposition 8.12. Taking into account Lemma 8.11, the result is easily generalized for Problem **C**. □

8.4. Uniqueness of blowups at contact points

The main result of this section is the uniqueness of blowups at contact points. As an immediate corollary this gives the existence of nontangential limits of second derivatives as we approach contact points.

Theorem 8.13 (Uniqueness of blowups at contact points). *Let $u \in P_1^+(0, M)$ be a solution of Problem* **A**, **B**, *or* **C**. *Then there exists a unique limit*

$$u_0(x) = \lim_{r \to 0+} u_r(x) = \lim_{r \to 0+} \frac{u(rx)}{r^2}.$$

In view of Lemma 8.11 and Corollary 8.6, it will be enough to prove the theorem only for Problems **A**, **B**. A slightly restated version of this theorem is as follows.

Lemma 8.14. *Let $u \in P_1^+(0, M)$ be a solution of Problem* **A** *or* **B** *and suppose that there exists a blowup u_0 of u at 0 such that*

$$u_0(x) = \frac{x_n^2}{2} + a_0 x_1 x_n. \tag{8.8}$$

Then all blowups of u at the origin have the same representation as in (8.8) *with the same constant a_0.*

Proof. First let us extend u across Π by odd reflection. Let $r_j \to 0+$ be such that $u_{r_j} = u(r_j x)/r_j^2$ converges, for a subsequence and in appropriate spaces, to a global solution

$$u_1(x) = \frac{1}{2}(x \cdot Ax) = \frac{1}{2} \sum a_{ij} x_i x_j,$$

where A is a symmetric matrix with entries a_{ij}. More precisely, $a_{ij} = 0$ for $i, j < n$, and $a_{nn} = 1$.

Now let $t_j \to 0+$ be another sequence with the respective rescalings u_{t_j}. By a similar argument, we obtain a limiting polynomial

$$u_2(x) = \frac{1}{2}(x \cdot Bx) = \frac{1}{2} \sum b_{ij} x_i x_j.$$

Here again B is a symmetric matrix with entries b_{ij} such that $b_{ij} = 0$ for $i, j < n$, and $b_{nn} = 1$.

Now for a directional vector $e \perp e_n$, consider the ACF monotonicity formula for $(\partial_e u)^{\pm}$ (see Exercise 8.8). Then $\phi_e(r, u) = \Phi(r, (\partial_e u)^{\pm})$ is a monotone nondecreasing function of r. Since ϕ_e is monotone, the limit as $r \to 0+$ exists and

$$\lim_{r\to 0+} \phi_e(r,u) = \lim_{r\to 0+} \phi_e(1,u_r) = C(n) a_0^4 (e\cdot e_1)^4. \tag{8.9}$$

Passing to the limit in (8.9) over the sequences $r = r_j \to 0+$ and $r = t_j \to 0+$, we obtain

$$\phi_e(1,u_1) = \phi_e(1,u_2). \tag{8.10}$$

In terms of the matrices A and B in the representations of u_1 and u_2, (8.10) is equivalent to having

$$|Ae| = |Be| = c(n)|a_0(e\cdot e_1)|, \tag{8.11}$$

for any $e \perp e_n$. We claim that this implies that $A = B$. Indeed, observe that by (8.11) we have the representations

$$u_1 = \frac{x_n^2}{2} + a x_1 x_n, \quad u_2 = \frac{x_n^2}{2} + b x_1 x_n, \quad \text{with } |a| = |b| = |a_0|.$$

If $a_0 = 0$, then we are done. Otherwise, to prove that $a = b$ it will suffice to show that $\{u_1 < 0\} \equiv \{u_2 < 0\}$. Assuming that these sets are not equal, we can find a point $e^0 \in (\partial B_1) \cap \mathbb{R}^n_+$ such that $u_1(e^0) < 0$ but $u_2(e^0) > 0$. Consider the line segment $\ell_0 = [0, e^0] = \{te^0 : 0 \le t \le 1\}$. Then we claim that there exists $r_0 > 0$ such that u does not change sign on $\ell_0 \cap B_{r_0}$. Once we show the latter statement, we will be done. So suppose it fails. Then there exists a sequence $\rho_j \to 0+$ such that $u(\rho_j e^0) = 0$. Consider then the rescalings $u_{\rho_j}(x) = u(\rho_j x)/\rho_j^2$. Applying the ACF monotonicity formula (after the odd reflection of u) as above, we end up with a limit function (global solution) which has the representation

$$u_3(x) = \frac{x_n^2}{2} + c x_1 x_n.$$

Moreover, the above analysis gives that $|c| = |a|$. Hence $c = a$ or $c = b$, i.e., $u_3 = u_1$ or $u_3 = u_2$. In particular $u_3(e^0) \neq 0$. On the other hand, by our construction we have $u_{\rho_j}(e^0) = 0$, which implies that $u_3(e^0) = 0$. This is clearly a contradiction. □

Proof of Theorem 8.13. The proof is a simple combination of Corollary 8.6 and Lemmas 8.11 and 8.14. □

The uniqueness of blowups can also be seen as a statement on the existence of nontangential limits of second derivatives.

Theorem 8.15. *Let $u \in P_1^+(0, M)$ be a solution of Problem* **A**, **B**, *or* **C**. *Then the limits*

$$\lim_{x \to 0} \partial_{x_i x_j} u, \quad i, j = 1, \dots, n,$$

exist nontangentially for $x \in \Omega(u)$.

Proof. Let $x^j \to 0$ nontangentially, i.e. $x_n^j \geq \kappa |x^j|$ for some $\kappa > 0$. Let

$$r_j = |x^j|, \quad \xi^j = \frac{x^j}{r_j} \in \partial B_1, \quad u_j(x) = \frac{u(r_j x)}{r_j^2}.$$

Obviously $\xi_n^j \geq \kappa$, $u_j \to u_0$ in $C^{1,\alpha}_{\mathrm{loc}}(\mathbb{R}^n_+ \cup \Pi)$, and $\xi^j \to \xi^0$ for a subsequence. Since by Proposition 8.9 we have $B_\kappa(\xi^j) \subset \Omega(u_j)$, it follows that $u_j \to u_0$ in $C^2(B_{\kappa/2}(\xi^0))$. In particular,

$$D^2 u(x^j) = D^2 u_j(\xi^j) \to D^2 u_0(\xi^0).$$

Since $u_0 \in P_\infty^+(0, M)$, by Corollary 8.6 and Lemma 8.14, $D^2 u_0(\xi^0)$ is uniquely defined. □

Notes

The behavior of the free boundary near the contact points with the fixed boundary in the classical obstacle problem was studied by Apushkinskaya-Uraltseva [**AU95**], where the tangential touch was proved. The C^1 touch was established by Uraltseva [**Ura96**]. These results were obtained for solutions vanishing on the fixed boundary. In Problem **A**, the C^1 character of the touch was proved by Shahgholian-Uraltseva [**SU03**] (for solutions vanishing on a C^3 fixed boundary). The nature of the contact is more complicated when the solution does not vanish on the fixed boundary. For the classical obstacle problem, the contact with nonhomogeneous Dirichlet data on the fixed boundary was studied by Andersson-Shahgholian [**AS05**]. For Problem **A**, the contact when the solution is $C^{2,\mathrm{Dini}}$ on a flat fixed boundary was studied by Andersson [**And07**]. A similar study for Problem **C** was done by Andersson-Matevosyan-Mikayelyan [**AMM06**] and Andersson-Mikayelyan [**AM12**]. Uniform $C^{1,1}$ estimates up to the fixed boundary in that problem were proved by Apushkinskaya-Uraltseva [**AU07**]. In Problem **B**, the tangential touch was proved by Matevosyan [**Mat05**]; however, no uniform C^1 estimates for the free boundary are known near the fixed one, unlike in Problems **A** and **C**.

For the parabolic counterpart of the results in this section for Problem **A**, we refer to the papers of Apushkinskaya-Shahgholian-Uraltseva [**ASU00, AUS02, AUS03**] and Apushkinskaya-Matevosyan-Uraltseva [**AMU09**]. For Problem **C**, see Apushkinskaya-Uraltseva [**AU09**].

Exercises

8.1. State and prove Weiss's monotonicity formula for solutions in half-balls B_R^+. In particular prove (8.5).

8.2. Using Weiss's monotonicity formula in Exercise 8.1 one can give a simpler proof of Theorem 8.5 in the case of Problem **A**. Use the following strategy, if the origin is a contact point:

a) Prove that solutions homogeneous of degree 2 are polynomials $p(x)$ and that $W(r,p)\equiv c_0$ with c_0 independent of p, as long as $\Delta p = 1$.

b) Prove that u_∞ and u_0 are homogeneous of degree 2.

c) Use that $c_0 = W(0+,u)\leq W(r,u)\leq W(\infty,u)=c_0$ to conclude that u is also homogeneous of degree 2 and thus is a polynomial by a).

8.3. Let a function u defined in $\mathbb{R}^n$ satisfy the following condition: the directional derivative $\partial_e u$ does not change sign in $\mathbb{R}^n$ for any direction $e\perp e_n$. Then $u=u(x_1,x_n)$ in some rotated coordinate system preserving the direction e_n.

8.4. Let u be a solution of Problem **A** or **B** in B_r^+, vanishing on B_r'. Prove the following nondegeneracy properties: (i) if $x^0\in\overline{\Omega(u)\cap\{u>0\}}$, then

$$\sup_{B_\rho(x^0)\cap\mathbb{R}^n_+} u \geq u(x^0)+\frac{\rho^2}{2n},$$

provided $B_\rho(x^0)\subset B_r$; (ii) if $0\in\Gamma'(u)$, then in any neighborhood of the origin there are points where u is positive.

8.5. Let u be a solution of Problem **C** in B_r^+, vanishing on B_r'. Prove the following nondegeneracy property:

$$\sup_{B_\rho(x^0)\cap\mathbb{R}^n_+} u \geq u(x^0)+c_0\rho^2 \quad \text{if } x^0\in\overline{\Omega_+(u)},$$

$$\sup_{B_\rho(x^0)\cap\mathbb{R}^n_+} (-u) \geq -u(x^0)+c_0\rho^2 \quad \text{if } x^0\in\overline{\Omega_-(u)},$$

provided $B_\rho(x^0)\subset B_r$, with $c_0=c_0(n,\lambda_\pm)>0$.

8.6. Let u be a solution of Problem **A** in $B_2(e_n)\cap\mathbb{R}^n_+$, vanishing on $B_2(e_n)\cap\Pi$. Prove that if

$$\sup_{B_2(e_n)\cap\mathbb{R}^n_+} |u|\leq\varepsilon<\frac{1}{16n},$$

then $u\equiv 0$ in $Q:=(e_n-\mathcal{C}_1)\cap\mathbb{R}^n_+$.

Proof. 1) If there is $y \in Q$ with $u(y) > 0$, then by nondegeneracy

$$\sup_{B_\rho(y)\cap\mathbb{R}^n_+} u \geq u(y) + \frac{\rho^2}{2n} > \frac{\rho^2}{2n}$$

if $B_\rho(y) \subset B_2(e_n)$. Then $\rho^2 < 2n\varepsilon$, implying that $\rho < 1/\sqrt{8}$. On the other hand, for any $y \in Q$, $B_{\rho_0}(y) \subset B_2(e_n)$ with $\rho_0 = 2 - \sqrt{2} > 1/\sqrt{8}$, a contradiction. Thus, $u \leq 0$ in Q

2) If $u \leq 0$ in Q and $u = 0$ somewhere in Q, then $u \equiv 0$.

3) If $u < 0$ in Q, then $\Delta u = 1$ in Q and one can apply the interior nondegeneracy (Lemma 3.3) for arbitrary $y \in Q$. Then

$$\sup_{B_\rho(y)} u \geq u(y) + \frac{\rho^2}{2n}$$

if $B_\rho(y) \subset B_2(e_n) \cap \mathbb{R}^n_+$, which implies that $\rho^2 < 4n\varepsilon < 1/4$. Now choose $y = (1-\delta)e_n$ with a small $\delta > 0$ and let $\rho = 1 - \delta$. Then $B_\rho(y) \subset B_2(e_n) \cap \mathbb{R}^n_+$, which by the said above should imply that $(1-\delta)^2 < 1/4$. But this is clearly a contradiction if δ is small enough. Thus only the case 2) is possible, i.e. $u \equiv 0$ in Q. □

8.7. Prove Lemma 8.11 with the assumption $u \in P_1^+(M)$ instead of $u \in P_1^+(0, M)$.

Hint: If in any neighborhood of $x = 0$ there are points of both phases $\Omega_\pm$, then $|\nabla u(0)| = 0$.

8.8. Let $v \in C(B_R^+ \cup B_R')$ be such that $v = 0$ on B_R' and $v^\pm$ are subharmonic in B_R^+. Extend v by odd reflection to B_R^-. Then for the extended function v, the $v^\pm$ are subharmonic in B_R.

Chapter 9

The thin obstacle problem

In this chapter we study a variant of the obstacle problem where the obstacle is given only on a lower-dimensional (or "thin") set. More specifically, we consider a simplified model, which we call Problem **S** (see §9.1). In §9.2 we prove the $C^{1,\alpha}$ regularity of the solutions, by using the method of penalization. We then discuss Almgren's frequency formula (§9.3), which leads to a specific notion of rescalings and blowups, as well as a classification of free boundary points (§9.4). The study of homogeneous global solutions then leads to the optimal $C^{1,1/2}$ regularity of solutions (§9.5). We then turn to the study of the free boundary. In §9.6 we prove the $C^{1,\alpha}$ regularity of the set of regular points where the blowups have the minimal homogeneity 3/2. In §9.7 we start the investigation of singular points, where the coincidence set has a zero H^{n-1} density. In §9.8 we prove analogues of Weiss's and Monneau's monotonicity formulas, which then provide us with another notion of blowup (so-called homogeneous blowup), which is more in line with the classical obstacle problem. The uniqueness and continuous dependence of such blowups then results in a structural theorem for the singular set (§9.9).

9.1. The thin obstacle problem

Let D be a domain in $\mathbb{R}^n$ and $\mathcal{M}$ a smooth $(n-1)$-dimensional manifold in $\mathbb{R}^n$ that divides D into two parts: D_+ and D_-. For given functions $\psi : \mathcal{M} \to \mathbb{R}$ and $g : \partial D \to \mathbb{R}$ satisfying $g > \psi$ on $\mathcal{M} \cap \partial D$, consider the problem of minimizing the Dirichlet integral

$$J(u) = \int_D |\nabla u|^2 dx,$$

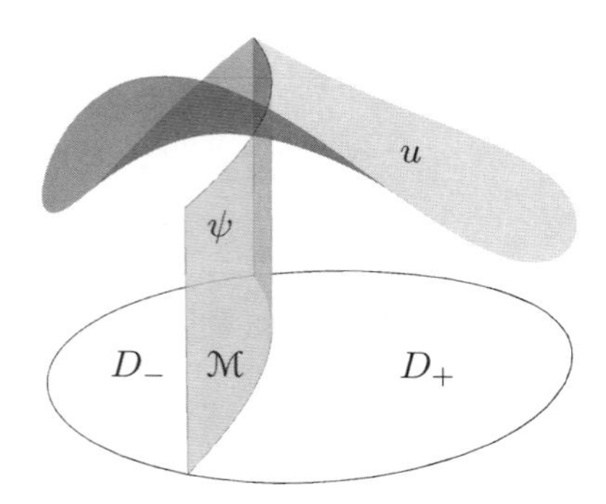

Figure 9.1. The thin obstacle (Signorini) problem

on the closed convex set

$$\mathfrak{K}_{g,\psi,\mathcal{M}} = \{u \in W^{1,2}(D) : u = g \text{ on } \partial D,\ u \geq \psi \text{ on } \mathcal{M} \cap D\}.$$

This problem is known as the *thin obstacle problem*, with ψ known as the *thin obstacle*. The main difference from the classical obstacle problem is that u is constrained to stay above the obstacle ψ only on $\mathcal{M}$ and not on the entire domain D (see Fig. 9.1).

When $\mathcal{M}$ and ϕ are smooth, it is known that the minimizer u in the thin obstacle problem is of class $C^{1,\alpha}_{\mathrm{loc}}(D_\pm \cup \mathcal{M})$ (see [**Caf79**] and [**Ura85**]; see also Theorem 9.2 below). Since we can make free perturbations away from $\mathcal{M}$, it is easy to see that u satisfies

$$\Delta u = 0 \quad \text{in } D \setminus \mathcal{M} = D_+ \cup D_-,$$

but in general u does not need to be harmonic across $\mathcal{M}$. Instead, on $\mathcal{M}$, one has the following *complementarity conditions*:

$$u - \psi \geq 0, \quad \partial_{\nu^+} u + \partial_{\nu^-} u \geq 0, \quad (u - \psi)(\partial_{\nu^+} u + \partial_{\nu^-} u) = 0,$$

where $\nu^\pm$ are the outer unit normals to $D_\pm$ on $\mathcal{M}$. One of the main objects of study in this problem is the *coincidence set*

$$\Lambda(u) := \{x \in \mathcal{M} : u(x) = \psi(x)\}$$

and the free boundary (in the relative topology on $\mathcal{M}$)

$$\Gamma(u) := \partial_{\mathcal{M}} \Lambda(u).$$

$\Gamma(u)$ is sometimes called a *thin* (or *lower-dimensional*) *free boundary*, since it has a codimension greater than 1.

A similar problem is obtained when $\mathcal{M}$ is a part of ∂D and one minimizes $J(u)$ over the convex set

$$\mathfrak{K} = \{u \in W^{1,2}(D) : u = g \text{ on } \partial D \setminus \mathcal{M},\ u \geq \psi \text{ on } \mathcal{M}\}.$$

In this case u is harmonic in D and satisfies the so-called *Signorini boundary conditions*

$$u - \psi \geq 0, \quad \partial_\nu u \geq 0, \quad (u - \psi)\partial_\nu u = 0$$

on $\mathcal{M}$, where ν is the outer unit normal on ∂D. This problem is known as the *boundary thin obstacle problem* or the *(scalar) Signorini problem*. Note that in the case when $\mathcal{M}$ is a plane and D and g are symmetric with respect to $\mathcal{M}$, the thin obstacle problem in D is equivalent to the boundary obstacle problem in D_+.

9.1.1. Application: semipermeable membranes. The thin obstacle problem arises in a variety of situations of interest in the applied sciences. Here we would like to describe an application relevant to math biology.

A semipermeable membrane is a membrane that is permeable only for a certain type of molecules (solvents) and blocks other molecules (solutes). Because of the chemical imbalance, the solvent flows through the membrane from the region of smaller concentration of solute to the region of higher concentration (osmotic pressure). The flow occurs in one direction and continues until a sufficient pressure builds up on the other side of the membrane (to compensate for osmotic pressure), which then shuts the flow. This process is known as osmosis.

Mathematically, let D represent the region occupied with a chemical solution and let $\mathcal{M} \subset \partial D$ be the semipermeable part of the boundary. Further, let u be the pressure of the chemical solution, which we assume to be a slightly compressible fluid. In the stationary case it satisfies

$$\Delta u = 0 \quad \text{in } D.$$

Let $\psi : \mathcal{M} \to \mathbb{R}$ be a given function that represents the osmotic pressure. Then on $\mathcal{M}$ we have the following boundary conditions:

$$\begin{aligned} u > \psi \quad &\Rightarrow \quad \partial_\nu u = 0 \qquad &&\text{(no flow)},\\ u \leq \psi \quad &\Rightarrow \quad \partial_\nu u = \lambda(u - \psi) \qquad &&\text{(flow)}, \end{aligned}$$

where λ is the permeability constant (finite permeability). Letting $\lambda \to \infty$ (infinite permeability), we obtain the Signorini boundary conditions on $\mathcal{M}$,

$$u \geq \psi, \quad \partial_\nu u \geq 0, \quad (u - \psi)\partial_\nu u = 0.$$

We refer to the book by Duvaut-Lions [**DL76**, Chapter I, §2.2], for details.

9.1.2. Application: optimal pricing. Motivated by modeling the American options for stocks with possible discontinuities, we consider a version of the optimal stopping problem in §1.1.8 for jump processes. Let $\mathbf{X}_x$ be an α-stable Lévy process ($0 < \alpha < 2$) with $\mathbf{X}_x(0) = x$ that will model the logarithm of the stock price. Let $g : \mathbb{R}^n \to \mathbb{R}$ be a payoff function in the sense that we generate a profit of $g(s)$ when trading the stock at $\mathbf{X}_x = s$. We then want to maximize the expected profit

$$u(x) = \sup_{\theta<\infty \text{ stopping time}} \mathbb{E}\left[g(\mathbf{X}_x(\theta))\right].$$

In mathematical finance, this problem is known as *perpetual American options.* A simple analysis shows that the value function u solves an obstacle-type problem

$$u \geq g, \quad (-\Delta)^{\alpha/2}u \geq 0, \quad (u-g)((-\Delta)^{\alpha/2}u) = 0 \quad \text{in } \mathbb{R}^n.$$

Here $(-\Delta)^{\alpha/2}$ is the so-called fractional Laplacian, which is a nonlocal integro-differential operator, defined for instance by the Fourier transform as

$$\widehat{(-\Delta)^{\alpha/2}u} = |\xi|^\alpha \widehat{u}(\xi).$$

In the case when $\alpha = 1$, this problem has a direct connection to the thin obstacle problem. Indeed, by adding an extra dimension and extending u harmonically to $\mathbb{R}^{n+1}_+ = \mathbb{R}^n \times (0,\infty)$ by solving the Cauchy problem

$$\Delta\tilde{u} = 0 \quad \text{in } \mathbb{R}^{n+1}_+, \quad \tilde{u}(\cdot, 0) = u,$$

we obtain a classical result that $-\partial_{x_{n+1}}\tilde{u}(x,0) = C_n(-\Delta)^{1/2}u(x)$ and thus $\tilde{u}$ solves a boundary thin obstacle problem in $\mathbb{R}^{n+1}_+$. While the problem for u is nonlocal, one effectively localizes the problem by adding an extra dimension. This extension method has been recently generalized for all $0 < \alpha < 2$ by Caffarelli-Silvestre [**CS07**], which made it possible to extend many results for the thin obstacle problem to the ones for the fractional Laplacian for any $0 < \alpha < 2$; see Caffarelli-Salsa-Silvestre [**CSS08**]. We also refer to Silvestre [**Sil07**] for the study of the problem by "intrinsic" methods, i.e., without using the extensions.

9.1.3. Model Problem S. We now describe the model problem that we will study in detail in the rest of this chapter. In short, it is the Signorini problem on B^+_R with zero thin obstacle on B'_R.

Problem S. We say that u is a solution of Problem **S** in B^+_R if $u \in W^{1,2}(B^+_R)$, $u \geq 0$ a.e. on B'_R (in the sense of traces), and u minimizes the Dirichlet integral

$$J(v) = \int_{B^+_R} |\nabla v|^2$$

among all functions in $\mathfrak{K} = \{v \in W^{1,2}(B_R^+) : v|_{B'_R} \geq 0,\ v|_{(\partial B_R)^+} = u\}$.

Assuming the $C^{1,\alpha}$ regularity of u on $B_R^+ \cup B'_R$, which we prove in the next section (see Theorem 9.2), the problem can be reformulated as

$$(\mathbf{S})\qquad \begin{aligned} \Delta u = 0 \quad &\text{in } B_R^+,\\ u \geq 0, \quad -\partial_{x_n} u \geq 0, \quad u\,\partial_{x_n} u = 0 \quad &\text{on } B'_R; \end{aligned}$$

see Exercise 9.1. As before, we denote by $\Lambda(u) = \{u(\cdot, 0) = 0\}$ the coincidence set and by $\Gamma(u) = \partial\Lambda \cap B'_R$ the free boundary.

In what follows we will automatically extend u to the entire ball B_R by even symmetry in x_n. Then, the resulting function will satisfy

$$\Delta u = 0 \quad \text{on } B_R \setminus \Lambda(u).$$

We may also write

$$\Delta u = 2(\partial_{x_n} u) H^{n-1}\big|_{\Lambda(u)} \quad \text{in } B_R$$

in the sense of distributions. The condition $u\partial_{x_n} u = 0$ on B'_R is then equivalent to

$$u\Delta u = 0 \quad \text{in } B_R.$$

9.2. $C^{1,\alpha}$ regularity

Let u be a solution of Problem $\mathbf{S}$ in B_1^+. In this section we will use a penalization method to prove the $C^{1,\alpha}$ regularity of u in $B_1^\pm \cup B'_1$. For small $\varepsilon > 0$ consider the solution u^ε of the *penalized* Problem $\mathbf{S}_\varepsilon$:

$$(\mathbf{S}_\varepsilon)\qquad \begin{aligned} \Delta u^\varepsilon = 0 \quad &\text{in } B_1^+,\\ -u^\varepsilon_{x_n} + \beta_\varepsilon(u^\varepsilon) = 0 \quad &\text{on } B'_1,\\ u^\varepsilon = u \quad &\text{on } (\partial B_1)^+, \end{aligned}$$

where $\beta_\varepsilon \in C^\infty(\mathbb{R})$ is such that

$$\beta_\varepsilon \leq 0, \quad \beta'_\varepsilon \geq 0, \quad \beta_\varepsilon(s) = 0 \text{ for } s \geq 0, \quad \beta_\varepsilon(s) = \varepsilon + s/\varepsilon \text{ for } s \leq -2\varepsilon^2.$$

The solution of the above problem is of class $C^\infty(B_1^+ \cup B'_1)$ (see e.g. Agmon-Douglis-Nirenberg [**ADN59**]) and satisfies the integral identity

$$\int_{B_1^+} \nabla u^\varepsilon \nabla \eta = -\int_{B'_1} \beta_\varepsilon(u^\varepsilon)\eta, \tag{9.1}$$

for any $\eta \in C_0^\infty(B_1)$. Arguing as in the proof of the energy inequality, it is easy to show that the family $\{u^\varepsilon\}$ is uniformly bounded in $W^{1,2}(B_1^+)$ and converges (weakly) to the function u (see Exercise 9.2). In fact, by carefully choosing the test functions η one may extract much more from (9.1).

Lemma 9.1 (Filling holes). *Let u be a solution of Problem* **S** *in B_1^+. Then*

$$\int_{B_\rho(x^0)} \frac{|\nabla(\partial_{x_k} u)|^2}{|x-x^0|^{n-2}} \le \frac{C_n}{\rho^n} \int_{B_{2\rho}(x^0)\setminus B_\rho(x^0)} (\partial_{x_k} u)^2, \tag{9.2}$$

for any $x^0 \in B'_{1/2}$, $k = 1, \dots, n$, and $\rho \in (0, 1/4)$, where C_n depends only on the dimension. In particular, $u \in W^{2,2}(B^+_{3/4})$.

Proof. See Exercise 9.3. □

With the above lemma at hand, we can now prove the $C^{1,\alpha}$ regularity of u.

Theorem 9.2. *Let u be a solution of Problem* **S** *in B_R^+. Then there exists $\alpha > 0$ such that $u \in C^{1,\alpha}_{\mathrm{loc}}(B_R^+ \cup B'_R)$. Moreover*

$$\|u\|_{C^{1,\alpha}(K)} \le C \|u\|_{L^2(B_R^+)},$$

for any $K \Subset B_1^+ \cup B'_1$, where $C = C(n, \operatorname{dist}(K, (\partial B_R)^+))$.

Proof. Without loss of generality we will assume that $R = 1$ and prove only that $u \in C^{1,\alpha}(B^+_{1/2} \cup B'_{1/2})$. By the classical results for the solutions of elliptic equations with Neumann boundary conditions, it is enough to establish that $\partial_{x_n} u$ is Hölder continuous on $B'_{1/2}$. In turn, this will follow once we show that for some $\alpha > 0$,

$$\int_{B_\rho^+(x^0)} \frac{|\nabla(\partial_{x_n} u)|^2}{|x-x^0|^{n-2}} dx \le C\,\rho^\alpha \tag{9.3}$$

for any $x^0 \in B'_{1/2}$ and $0 < \rho < 1/4$; see Exercise 9.4. To this end, fix $x^0 \in B'_{1/2}$ and define

$$I^{(k)}(\rho) := \int_{B_\rho^+(x^0)} \frac{|\nabla(\partial_{x_k} u)|^2}{|x-x^0|^{n-2}} dx, \quad k = 1, \dots, n.$$

We then have the following lemma.

Lemma 9.3. *There exists $\theta_n < 1$ depending only on the dimension such that for any $0 < \rho \le 1/4$ either one or the other of the following holds:*

$$I^{(n)}(\rho) \le \theta_n I^{(n)}(2\rho),$$
$$I^{(i)}(\rho) \le \theta_n I^{(i)}(2\rho), \quad \textit{for all } i = 1, \dots, n-1.$$

Proof. The proof is a combination of estimate (9.2) and the Poincaré inequality together with the Signorini boundary conditions. Since we already know that $u \in W^{2,2}(B^+_{3/4})$ by Lemma 9.1, we will have that $u\,\partial_{x_n} u = 0$ a.e.

on $B'_{3/4}$ (see Exercise 9.1). Then either one or the other of the following holds:

(i) $H^{n-1}(\{\partial_{x_n} u = 0\} \cap [B'_{2\rho}(x^0) \setminus B'_\rho(x^0)]) \geq c_n \rho^{n-1}$;

(ii) $H^{n-1}(\{u = 0\} \cap [B'_{2\rho}(x^0) \setminus B'_\rho(x^0)]) \geq c_n \rho^{n-1}$.

In case (i), by the Poincaré inequality we may estimate the right-hand side in (9.2) for $k = n$ by $\rho^{2-n} \int_{B^+_{2\rho}(x^0)\setminus B^+_\rho(x^0)} |\nabla(\partial_{x_n} u)|^2$ and arrive at

$$I^{(n)}(\rho) \leq C_n[I^{(n)}(2\rho) - I^{(n)}(\rho)],$$

or equivalently,

$$I^{(n)}(\rho) \leq \theta_n I^{(n)}(2\rho), \quad \theta_n = \frac{C_n}{C_n + 1} < 1.$$

In case (ii), using that $\partial_{x_i} u = 0$ a.e. on $\{u(\cdot, 0) = 0\}$, for $i = 1, \dots, n-1$, we have that

$$H^{n-1}(\{\partial_{x_i} u = 0\} \cap [B'_{2\rho}(x^0) \setminus B'_\rho(x^0)]) \geq c_n \rho^{n-1},$$

for all $i = 1, \dots, n-1$. Arguing as for $\partial_{x_n} u$, we then conclude that

$$I^{(i)}(\rho) \leq \theta_n I^{(i)}(2\rho).$$

□

Now, to complete the proof of Theorem 9.2, we iterate the inequalities in Lemma 9.3 for $\rho_k = 1/2^{(k+2)}$, $k = 0, 1, \dots$, to obtain that for any $\rho \in (0, 1/4)$ either

$$I^{(n)}(\rho) \leq C\rho^\alpha I^{(n)}(1/4)$$

or

$$I^{(i)}(\rho) \leq C\rho^\alpha I^{(i)}(1/4) \quad \text{for all } i = 1, \dots, n-1,$$

where $\alpha = (1/2)\log_2 \theta_n$. In the latter case, expressing $\partial_{x_n x_n} u$ in terms of $\partial_{x_i x_i} u$ by the harmonicity of u in B^+_1, we conclude that

$$I^{(n)}(\rho) \leq \sum_{i=1}^{n-1} I^{(i)}(\rho) \leq C\rho^\alpha.$$

So in all the cases, we obtain that

$$I^{(n)}(\rho) \leq C\rho^\alpha,$$

which completes our proof by using Exercise 9.4. □

9.3. Almgren's frequency formula

One of the important tools in our study of the thin obstacle problem will be Almgren's theorem on monotonicity of the frequency [**Alm00**].

9.3.1. Harmonic functions. For a harmonic function u in the ball B_R consider the following quantity:

$$N(r) = N(r,u) = \frac{r\int_{B_r}|\nabla u|^2}{\int_{\partial B_r} u^2}, \quad 0<r<R,$$

which will be called *Almgren's frequency* (the justification for the name will be given in what follows). A theorem of Almgren then says that $N(r)$ is monotone nondecreasing. To show this, let us introduce

$$H(r) = \int_{\partial B_r} u^2, \quad 0<r<R.$$

Using spherical coordinates it is straightforward to see that

$$H'(r) = \frac{n-1}{r}H(r) + 2\int_{\partial B_r}(\partial_\nu u)u.$$

Then applying the divergence theorem to the last term and using that $\Delta(u^2) = 2|\nabla u|^2$ in B_R, we arrive at

$$H'(r) = \frac{n-1}{r}H(r) + 2\int_{B_r}|\nabla u|^2.$$

Hence, we can write

$$\frac{rH'(r)}{H(r)} = n-1+2N(r).$$

Thus, the monotonicity of $N(r)$ is equivalent to the monotonicity of

$$\Phi(r) = r\frac{d}{dr}\log H(r),$$

which in turn is equivalent to showing the log-convexity of $H(r)$ in $\log r$ (which is a fancy name for the convexity of $\log H(e^t)$ in t).

Now, writing u in the form

$$u(x) = \sum_{k=0}^\infty f_k(x),$$

where the f_k are homogeneous harmonic polynomials of degree k, we have that

$$H(r) = \sum_{k=0}^\infty \int_{\partial B_1} f_k^2(\theta) r^{n-1+2k}d\theta = r^{n-1}\sum_{k=0}^\infty a_k r^{2k},$$

where $a_k = \int_{\partial B_1} f_k(\theta)^2 d\theta \geq 0$. Then by the Cauchy-Schwarz inequality we obtain that

$$H(\sqrt{r_1 r_2}) \leq \sqrt{H(r_1)}\sqrt{H(r_2)}, \quad 0<r_i<R,\ i=1,2,$$

which is the required convexity for H. This implies that $N(r)$ is indeed monotone. Moreover, the case of equality in the Cauchy-Schwarz implies that if $N(r_1) = N(r_2) = \kappa$ for $0<r_1<r_2<R$, then $N(r)\equiv\kappa$ and

consequently $H(r) \equiv Cr^{n-1+2\kappa}$ in the interval (r_1, r_2), which is possible only if $\kappa = k \in \mathbb{N} \cup \{0\}$ and

$$u(x) = f_k(x).$$

Now the name *frequency* comes from the fact that in dimension two, in polar coordinates

$$f_k = c_k r^k \cos(k\theta + \phi)$$

and $N(r)$ coincides with the frequency of $\cos(k\theta + \phi)$.

9.3.2. Thin obstacle problem. It turns out that Almgren's frequency formula is valid also for the solutions of Problem **S** (as was first observed by Athanasopoulos-Caffarelli-Salsa [**ACS08**]). Recall that we automatically extend the solutions of Problem **S** in B_R^+ to B_R by even symmetry in x_n.

Theorem 9.4 (Almgren's frequency formula). *Let u be a nonzero solution of Problem* **S** *in B_R^+; then the frequency of u,*

$$r \mapsto N(r) = N(r, u) = \frac{r \int_{B_r} |\nabla u|^2}{\int_{\partial B_r} u^2},$$

is nondecreasing for $0 < r < R$. Moreover, $N(r, u) \equiv \kappa$ for $0 < r < R$ if and only if u is homogeneous of degree κ in B_R, i.e.

$$x \cdot \nabla u - \kappa u = 0 \quad \text{in } B_R.$$

Proof. Consider the quantities

$$H(r) = \int_{\partial B_r} u^2, \quad D(r) = \int_{B_r} |\nabla u|^2. \tag{9.4}$$

Using the notation $u_\nu = \partial_\nu u$, where ν is the outward unit normal to ∂B_r, we have

$$H'(r) = \frac{n-1}{r} H(r) + 2 \int_{\partial B_r} u u_\nu. \tag{9.5}$$

On the other hand, using that $\Delta(u^2/2) = u\Delta u + |\nabla u|^2 = |\nabla u|^2$ and integrating by parts, we obtain

$$\int_{\partial B_r} u u_\nu = \int_{B_r} |\nabla u|^2 = D(r). \tag{9.6}$$

Further, to compute $D'(r)$ we use Rellich's formula

$$\int_{\partial B_r} |\nabla u|^2 = \frac{n-2}{r} \int_{B_r} |\nabla u|^2 + 2 \int_{\partial B_r} u_\nu^2 - \frac{2}{r} \int_{B_r} (x \cdot \nabla u) \Delta u.$$

We now claim that the last integral is zero, and moreover that $(x \cdot \nabla u)\Delta u = 0$ in the sense of measures. Since $\Delta u = 2u_{x_n} H^{n-1}\big|_\Lambda(u)$, it will follow from the fact that

$$(x \cdot \nabla u) u_{x_n} = 0 \quad \text{on } B_1'.$$

Indeed, $x\cdot\nabla u = 0$ on $\Lambda(u)$ because of the $C^{1,\alpha}$ regularity on B_1', and $u_{x_n} = 0$ on $B_1' \setminus \Lambda(u)$. Hence,

$$D'(r) = \frac{n-2}{r} D(r) + 2\int_{\partial B_r} u_\nu^2. \tag{9.7}$$

Thus, we have

$$\begin{aligned}\frac{N'(r)}{N(r)} &= \frac{1}{r} + \frac{D'(r)}{D(r)} - \frac{H'(r)}{H(r)}\\ &= \frac{1}{r} + \frac{n-2}{r} - \frac{n-1}{r} + 2\left\{\frac{\int_{\partial B_r} u_\nu^2}{\int_{\partial B_r} u u_\nu} - \frac{\int_{\partial B_r} u u_\nu}{\int_{\partial B_r} u^2}\right\} \geq 0.\end{aligned}$$

The last inequality is obtained from the Cauchy-Schwarz inequality and implies the monotonicity statement in the theorem. Analyzing the case of equality in Cauchy-Schwarz, we obtain the second part of the theorem. □

9.4. Rescalings and blowups

To study the behavior of u near the free boundary we again want to resort to the method of rescalings and blowups. Here we see an important difference of Problem **S** from Problems **A**, **B**, and **C**: whereas in the latter problems one has to scale quadratically to preserve the equation, in Problem **S**, any scaling of the type

$$u_r(x) = c(r)u(rx),$$

where $c(r) > 0$ is an arbitrary constant, will preserve the structure of the problem. Thus, on one hand we have a certain freedom in choosing $c(r)$, but on the other hand we would like to have a certain normalization property. The following choice works particularly well:

$$u_r(x) = \frac{u(rx)}{\left(\frac{1}{r^{n-1}}\int_{\partial B_r} u^2\right)^{1/2}}, \tag{9.8}$$

which is the scaling that corresponds to the normalization

$$\|u_r\|_{L^2(\partial B_1)} = 1. \tag{9.9}$$

Assuming $0 \in \Gamma(u)$, we then want to study the *blowups* of u at the origin, i.e. the limits of the rescalings u_r over subsequences $r = r_j \to 0+$.

It is not hard to see the existence of blowups. First, note that we have the following scaling property for Almgren's frequency function

$$N(\rho, u_r) = N(r\rho, u), \quad r > 0,\ \rho < R/r. \tag{9.10}$$

Then by the monotonicity of the frequency (Theorem 9.4) we have

$$\int_{B_1} |\nabla u_r|^2 = N(1, u_r) = N(r, u) \leq N(R, u),$$

for $r < R$. Combining this with (9.9), we will have a uniform bound for the family $\{u_r\}$ in $W^{1,2}(B_1)$. Thus, for a subsequence $r = r_j \to 0+$, we obtain

$$
\begin{aligned}
&u_{r_j} \to u_0 \quad \text{weakly in } W^{1,2}(B_1),\\
&u_{r_j} \to u_0 \quad \text{in } L^2(\partial B_1),\\
&u_{r_j} \to u_0 \quad \text{in } C^1_{\mathrm{loc}}(B_1^\pm \cup B_1'),
\end{aligned}
\tag{9.11}
$$

for a certain $u_0 \in W^{1,2}(B_1)$. The weak convergence in $W^{1,2}(B_1)$ and the strong convergence in $L^2(\partial B_1)$ are immediate. The third convergence follows from the uniform $C^{1,\alpha}_{\mathrm{loc}}$ estimates on u_r in $B_1^\pm \cup B_1'$ in terms of L^2-norm of u_r in B_1, provided by Theorem 9.2.

Proposition 9.5 (Homogeneity of blowups). *Let u be a solution of Problem* **S** *in B_R^+ with $0 \in \Gamma(u)$ and u_0 a blowup of u as described above. Then u_0 is a nonzero global solution of Problem* **S**, *homogeneous of degree $\kappa = N(0+, u)$.*

Remark 9.6. Hereafter, by a *global solution* of Problem **S** we understand a solution in B_R^+ for all $R > 0$.

Remark 9.7. We emphasize here that although the blowups at the origin might not be unique, as a consequence of Proposition 9.5 they all have the same homogeneity.

Proof. It is clear that u_0 is harmonic in $\mathbb{R}^n_\pm$. The fact that u_0 satisfies the Signorini boundary conditions on B_1' follows from the C^1_{loc} convergence of u_{r_j} to u_0 on $B_1^\pm \cup B_1'$. Once we prove the homogeneity of u_0 in $\mathbb{R}^n$, the boundary conditions will be satisfied automatically on the entire Π.

Next, we claim that $H(r, u_0) = \int_{\partial B_r} u_0^2 > 0$ for any $0 < r < 1$. Indeed, if $H(r_0, u_0) = 0$ for some $0 < r_0 < 1$, then from subharmonicity of $u_0^\pm$ it follows that $u_0 \equiv 0$ in B_{r_0}. By the unique continuation for the harmonic functions, we must have $u_0 \equiv 0$ in $B_1^\pm$ and consequently in B_1. On the other hand, we have strong convergence $u_{r_j} \to u_0$ in $L^2(\partial B_1)$, which combined with the equalities $\int_{\partial B_1} u_{r_j}^2 = 1$ implies that $\int_{\partial B_1} u_0^2 = 1$. This is clearly a contradiction.

Having now that $H(r, u_0) > 0$ for $0 < r < 1$ and using the C^1_{loc} convergence $u_{r_j} \to u_0$ in $B_1^\pm \cup B_1'$, we pass to the limit to obtain

$$
N(r, u_0) = \lim_{r_j \to 0+} N(r, u_{r_j}) = \lim_{r_j \to 0+} N(r r_j, u) = N(0+, u) = \kappa.
$$

This implies that $N(r, u_0)$ is constant for $0 < r < 1$. In view of the last part of Theorem 9.4 we conclude that u_0 is a homogeneous function in B_1, and by unique continuation, also in $\mathbb{R}^n$. □

9.4.1. Classification of free boundary points. Almgren's frequency functional can be defined at any point $x^0 \in \Gamma(u)$ by simply translating that point to the origin:

$$N(r,u,x^0) := \frac{r \int_{B_r(x^0)} |\nabla u|^2}{\int_{\partial B_r(x^0)} u^2},$$

where $r > 0$ is such that $B_r(x_0) \Subset D$. This enables us to give the following definitions.

Definition 9.8 (Classification of free boundary points)**.** For a solution u of Problem **S** and $\kappa \geq 0$ let

$$\Gamma_\kappa(u) := \{x^0 \in \Gamma(u) : N(0+,u,x^0) = \kappa\}.$$

Note that $\kappa = N(0+,u,x^0)$ is exactly the homogeneity of the blowups of u at x^0. So an important questions to ask is for what values of κ do we have homogeneous global solutions of Problem **S** of degree κ? Note that because of the $C^{1,\alpha}$ regularity in $\mathbb{R}^n_\pm$, we immediately have that $\kappa \geq 1+\alpha$. However, the general question of possible values of κ is still open, except in dimension $n = 2$. Using polar coordinates in dimension $n = 2$ one can show that the only homogeneous global solutions are:

$$\hat{u}_\kappa = C_\kappa \operatorname{Re}(x_1 + i|x_2|)^\kappa, \quad \kappa = 2m - 1/2,\ 2m,\ m \in \mathbb{N},$$

and

$$\hat{v}_\kappa = C_\kappa \operatorname{Im}(x_1 + i|x_2|)^\kappa, \quad \kappa = 2m+1,\ m \in \mathbb{N}.$$

However, it can be shown that the solutions $\hat{v}_\kappa$ can never appear as blowups at free boundary points (see Exercise 9.6). Therefore the range of possible values of κ for blowups in dimension $n = 2$ is

$$\kappa = 2m - 1/2,\ 2m,\ m \in \mathbb{N}.$$

(See Fig. 9.2 for the illustration of some of these solutions.)

In higher dimensions, very few results are known in this direction, however, the next result establishes the minimum possible value of κ and leads to the optimal regularity of solutions of Problem **S**.

Proposition 9.9. *Let u be a homogeneous global solution of Problem* **S** *of degree $\kappa \in (1,2)$. Then $\kappa = 3/2$ and*

$$u(x) = C_n \operatorname{Re}(x_1 + i|x_n|)^{3/2},$$

after a possible rotation in $\mathbb{R}^{n-1}$.

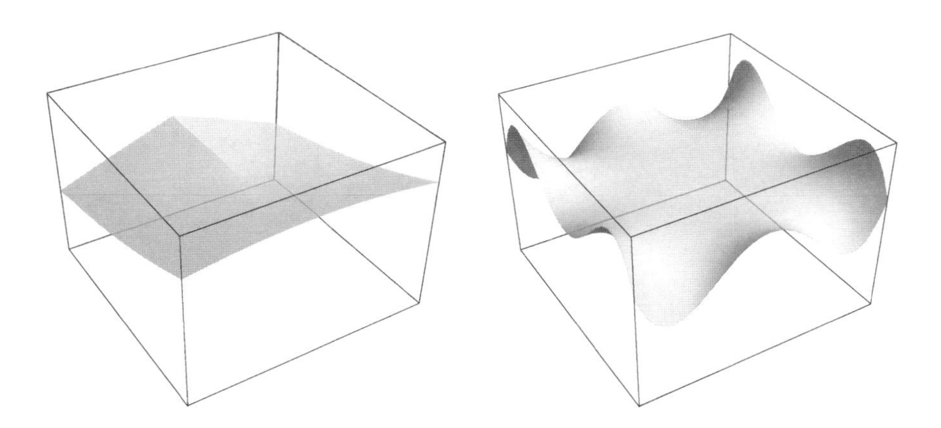

Figure 9.2. Graphs of $\operatorname{Re}(x_1 + i\,|x_2|)^{3/2}$ and $\operatorname{Re}(x_1 + i\,|x_2|)^6$

Proof. For a direction $e \in \partial B_1'$ consider the positive and negative parts of the tangential derivative $\partial_e u$. Then the following conditions are satisfied:

$$\Delta(\partial_e u)^\pm \geq 0, \quad (\partial_e u)^\pm \geq 0, \quad \partial_e u^+ \cdot \partial_e u^- = 0 \quad \text{in } \mathbb{R}^n;$$

see Exercise 9.5. Hence we can apply the ACF monotonicity formula to the pair $\partial_e u^\pm$. Namely, the functional

$$\phi_e(r, u) = \frac{1}{r^4} \int_{B_r} \frac{|\nabla(\partial_e u)^+|^2}{|x|^{n-2}} \int_{B_r} \frac{|\nabla(\partial_e u)^-|^2}{|x|^{n-2}}$$

is monotone nondecreasing in r. On the other hand, from the homogeneity of u, it is easy to see that

$$\phi_e(r, u) = r^{4(\kappa-2)}\phi_e(1, u), \quad r > 0.$$

Since $\kappa < 2$, $\phi_e(r)$ can be monotone increasing if and only if $\phi_e(1) = 0$ and consequently $\phi_e(r) = 0$ for all $r > 0$.

From this it follows that one of the functions $\partial_e u^\pm$ is identically zero, which is equivalent to $\partial_e u$ being either nonnegative or nonpositive on the entire $\mathbb{R}^n$. Since this is true for any tangential direction $e \in \partial B_1'$, it follows that u depends only on one tangential direction and is monotone in that direction (see Exercise 5.4). Therefore, without loss of generality we may assume that $n = 2$. However, we already know all possible homogeneous solutions in dimension $n = 2$, and the only one with $1 < \kappa < 2$ is

$$\hat{u}_{3/2}(x) = C_{3/2} \operatorname{Re}(x_1 + i|x_2|)^{3/2}.$$

This completes the proof of the proposition. □

Corollary 9.10 (Minimal homogeneity). *Let u be a solution of Problem* **S** *in B_R^+ and $0 \in \Gamma(u)$. Then*

$$N(0+, u) \geq \frac{3}{2}.$$

Moreover, either

$$N(0+, u) = \frac{3}{2} \quad \text{or} \quad N(0+, u) \geq 2.$$

In particular, we obtain that

$$\Gamma(u) = \Gamma_{3/2}(u) \cup \bigcup_{\kappa \geq 2} \Gamma_\kappa(u).$$

Besides Proposition 9.9, there is one more case when we can completely characterize the homogeneous global solutions.

Proposition 9.11. *Let u be a homogeneous global solution of Problem* **S** *of degree $\kappa = 2m$ for $m \in \mathbb{N}$. Then u is a homogeneous harmonic polynomial.*

Since $\Delta u = 2(\partial_{x_n} u) H^{n-1}|_{\Lambda(u)}$ on $\mathbb{R}^n$, with $\partial_{x_n} u \leq 0$ on Π, this is a particular case of the following lemma, which is essentially Lemma 7.6 in Monneau [**Mon09**], with an almost identical proof.

Lemma 9.12. *Let $v \in W^{1,2}_{\mathrm{loc}}(\mathbb{R}^n)$ satisfy $\Delta v \leq 0$ in $\mathbb{R}^n$ and $\Delta v = 0$ in $\mathbb{R}^n \setminus \Pi$. If v is homogeneous of degree $\kappa = 2m$, $m \in \mathbb{N}$, then $\Delta v = 0$ in $\mathbb{R}^n$.*

Proof. By assumption, $\mu := \Delta v$ is a nonpositive measure, living on Π. We are going to show that $\mu = 0$. To this end, let P be a $2m$-homogeneous harmonic polynomial, which is positive on $\Pi \setminus \{0\}$. For instance, take

$$P(x) = \sum_{j=1}^{n-1} \mathrm{Re}(x_j + i x_n)^{2m}.$$

Further, let $\psi \in C_0^\infty(0, \infty)$ with $\psi \geq 0$ and $\Psi(x) = \psi(|x|)$. Then we have

$$\begin{aligned}
-\langle \mu, \Psi P \rangle = -\langle \Delta v, \Psi P \rangle &= \int_{\mathbb{R}^n} \nabla v \cdot \nabla(\Psi P) \\
&= \int_{\mathbb{R}^n} \Psi \nabla v \cdot \nabla P + P \nabla v \cdot \nabla \Psi \\
&= \int_{\mathbb{R}^n} -\Psi v \Delta P - v \nabla \Psi \cdot \nabla P + P \nabla v \cdot \nabla \Psi \\
&= \int_{\mathbb{R}^n} -\Psi v \Delta P - v \frac{\psi'(|x|)}{|x|} (x \cdot \nabla P) + (x \cdot \nabla v) \frac{\psi'(|x|)}{|x|} P = 0,
\end{aligned}$$

where in the last step we have used that $\Delta P = 0$, $x \cdot \nabla P = 2mP$, $x \cdot \nabla v = 2mv$. This implies that the measure μ is supported at the origin. Hence

$\mu = c\delta_0$, where δ_0 is Dirac's delta. On the other hand, μ is $2(m-1)$-homogeneous and δ_0 is $(-n)$-homogeneous and therefore $\mu = 0$. $\square$

9.5. Optimal regularity

With the help of Proposition 9.9, we can now prove the optimal regularity of the solutions of Problem **S**.

Theorem 9.13 (Optimal regularity in the Signorini problem). *Let u be a solution of Problem* **S** *in B_R^+. Then $u \in C^{1,1/2}_{\mathrm{loc}}(B_R^+ \cup B'_R)$, and moreover*

$$\|u\|_{C^{1,1/2}(K)} \le C\|u\|_{L^2(B_R^+)},$$

for any $K \Subset B_R^\pm \cup B'_R$, where $C = C(n, \operatorname{dist}(K, (\partial B_R)^+))$.

Note that $C^{1,1/2}$ is indeed the best regularity possible, since the function

$$\hat{u}_{3/2}(x) = C_{3/2}\operatorname{Re}(x_1 + i|x_n|)^{3/2}$$

is a solution of Problem **S**.

The proof of Theorem 9.13 will follow from the growth estimate in the next lemma (with $\kappa = 3/2$) combined with the interior elliptic estimates.

Lemma 9.14 (Growth estimate). *Let u be a solution of Problem* **S** *in B_1^+, $0 \in \Gamma$, and $N(0+, u) \ge \kappa$. Then*

$$\sup_{B_r} |u| \le C_0\, r^\kappa, \quad 0 < r < 1/2,$$

with $C_0 = C(n, \kappa, \|u\|_{L^2(B_1)})$.

Proof. From the monotonicity of N we will have $N(r) \ge \kappa$ for all r. This is equivalent to having

$$r\frac{H'(r)}{H(r)} \ge n - 1 + 2\kappa.$$

Dividing by r and integrating from r to $3/4$ we will obtain

$$\log\frac{H(3/4)}{H(r)} \ge (n - 1 + 2\kappa)\log\frac{3/4}{r},$$

and consequently

$$H(r) \le C_0 r^{n-1+2\kappa}, \quad 0 < r < 3/4.$$

Next, observing that $u^\pm$ are subharmonic (see Exercise 9.5), we will obtain

$$\sup_{B_r} u^\pm \le C_n\Big(\frac{1}{r^{n-1}}\int_{\partial B_{(3/2)r}} (u^\pm)^2\Big)^{1/2} \le C_n\Big(\frac{H((3/2)r)}{r^{n-1}}\Big)^{1/2} \le C_0 r^\kappa,$$

which implies the desired estimate. $\square$

Proof of Theorem 9.13. Without loss of generality we assume that $R = 1$ and prove that $u \in C^{1,1/2}(B^+_{1/2} \cup B'_{1/2})$. For any $x \in B^+_{1/2}$ let

$$d(x) = \operatorname{dist}(x, \Gamma).$$

Note that $B_{d(x)}(x) \cap \{x_n = 0\}$ must be fully contained in either $\{u(\cdot, 0) = 0\}$ or $\{u(\cdot, 0) > 0\}$. Therefore, either the odd or the even reflection of u from B^+_1 to B^-_1 is going to be harmonic in $B_{d(x)}(x)$. We will denote the appropriate extension by $\tilde{u}$.

Now take two points $x^1, x^2 \in B^+_{1/2}$ with $|x^1 - x^2| \leq 1/8$. We want to show that

$$|\nabla u(x^1) - \nabla u(x^2)| \leq C|x^1 - x^2|^{1/2}, \tag{9.12}$$

with C depending on the L^2-norm of u in B_1.

1) Assume first that $d(x^1) \geq 1/4$ (or $d(x^2) \geq 1/4$). Then $\tilde{u}$ is harmonic in $B_{1/4}(x^1)$ and therefore (9.12) follows from the interior estimates for harmonic functions.

2) Suppose now that $d(x^2) \leq d(x^1) \leq 1/4$ and $|x^1 - x^2| \geq d(x^1)/2$. We then see that $\tilde{u}$ is harmonic in $B_{d(x^i)}(x^i)$, $i = 1, 2$, and from Lemma 9.14 we obtain that

$$|\tilde{u}| \leq C_0 d(x^i)^{3/2} \quad \text{in } B_{d(x^i)}(x^i).$$

By the interior gradient estimates we then have

$$|\nabla u(x^i)| = |\nabla \tilde{u}(x^i)| \leq C_0 d(x^i)^{1/2}.$$

Hence, in this case

$$|\nabla u(x^1) - \nabla u(x^2)| \leq |\nabla u(x^1)| + |\nabla u(x^2)| \leq C_0 d(x^1)^{1/2} \leq C_0|x^1 - x^2|^{1/2}.$$

3) Finally, suppose $d(x^2) \leq d(x^1) \leq 1/4$ and $|x^1 - x^2| \leq d(x^1)/2$. Then from harmonicity of $\tilde{u}$ in $B_{d(x^1)}(x^1)$ and the estimate $|\tilde{u}| \leq C_0 d(x_1)^{3/2}$, combined with the interior derivative estimates, we will have

$$|D^2\tilde{u}| \leq C_0 d(x_1)^{-1/2} \quad \text{in } B_{d(x^1)/2}(x^1)$$

and therefore

$$\begin{aligned}|\nabla u(x^1) - \nabla u(x^2)| = |\nabla \tilde{u}(x^1) - \nabla \tilde{u}(x^2)| &\leq C_0 d(x_1)^{-1/2}|x^1 - x^2| \\ &\leq C_0|x^1 - x^2|^{1/2}.\end{aligned}$$

This completes the proof of the theorem. □

9.6. The regular set

Recall that by Corollary 9.10, we can represent the free boundary $\Gamma(u)$ as the union

$$\Gamma(u) = \Gamma_{3/2}(u) \cup \bigcup_{\kappa \geq 2} \Gamma_\kappa(u).$$

It turns out that the points of the set $\Gamma_{3/2}(u)$ have many similarities with the regular points in the classical obstacle problem, which motivates the next definition.

Definition 9.15 (Regular points). For a solution u of Problem **S** we call the set $\Gamma_{3/2}(u)$ the *regular set* and the points $x^0 \in \Gamma_{3/2}(u)$ *regular points.*

We immediately observe the following.

Proposition 9.16. *For a solution u of Problem* **S**, *the set $\Gamma_{3/2}(u)$ is a relatively open subset of $\Gamma(u)$.*

Proof. Note that from Almgren's frequency formula it follows that the mapping $x^0 \mapsto N(0+, u, x^0)$ is upper semicontinuous on $\Gamma(u)$. Moreover, since $N(0+, u, x^0)$ does not take values in the interval $(3/2, 2)$, one immediately obtains that $\Gamma_{3/2}(u)$ is a relatively open subset of $\Gamma(u)$. □

The main result of this section is the following theorem.

Theorem 9.17 (Regularity of the regular set). *Let u be a solution of Problem* **S**. *Then the regular set $\Gamma_{3/2}(u)$ is locally a $C^{1,\alpha}$ regular $(n-2)$-dimensional surface.* □

Similarly to the classical obstacle problem (or Problem **A**), we split Theorem 9.17 into two parts.

Proposition 9.18 (Lipschitz regularity). *Let u be a solution of Problem* **S** *in B_1 with $0 \in \Gamma_{3/2}(u)$. Then there exists $\rho = \rho_u > 0$ such that*

$$\Gamma(u) \cap B'_\rho = \Gamma_{3/2}(u) \cap B'_\rho = \{x_{n-1} = g(x'') : x'' \in B''_\rho\},$$

after a possible rotation in $\mathbb{R}^{n-1}$, where $g \in C^{0,1}(B''_\rho)$.

Proposition 9.19 (Lipschitz implies $C^{1,\alpha}$). *Let u, $\rho = \rho_u$, and let the function g be as in Proposition* 9.18. *Then $g \in C^{1,\alpha}(B''_\rho)$.*

The proof of the Lipschitz regularity of the free boundary is based on the idea of the directional monotonicity that we used in Problems **A**, **B**, and **C**. If we consider a blowup u_0 at a point $x^0 \in \Gamma_{3/2}(u)$, then after a rotation in $\mathbb{R}^{n-1}$ we may write it in the form

$$u_0(x) = C_0 \operatorname{Re}(x_{n-1} + i|x_n|)^{3/2}.$$

It is then easy to calculate that for any $e \in \partial B_1'$,

$$\begin{aligned}\partial_e u_0(x) &= \frac{3}{2} C_0 (e \cdot e_{n-1}) \operatorname{Re}(x_{n-1} + i|x_n|)^{1/2} \\ &= \frac{3}{2\sqrt{2}} C_0 (e \cdot e_{n-1}) \sqrt{\sqrt{x_{n-1}^2 + x_n^2} + x_{n-1}}.\end{aligned}$$

We thus have the following properties for $\partial_e u_0$: for any $\eta > 0$ and $c_0 > 0$, there exists $\delta_0 = \delta_0(\eta, c_0) > 0$ such that for any $e \in \mathcal{C}_\eta'$,

$$\begin{aligned}\partial_e u_0(x) &\geq 0, \quad x \in B_1, \\ \partial_e u_0(x', x_n) &\geq 2\delta_0 > 0, \quad x \in B_1 \cap \{|x_n| \geq c_0\}.\end{aligned}$$

The next lemma essentially says that these estimates hold also for C^1 approximations of the global solution u_0.

Lemma 9.20. *Let h be a continuous function in B_1 and Λ a certain closed subset of $\mathbb{R}^{n-1}$. Then for any $\delta_0 > 0$ there exists $\varepsilon_0 = \varepsilon_0(n, \delta_0) > 0$ such that if*

(i) *$h = 0$ on $B_1 \cap \Lambda$,*
(ii) *$\Delta h \leq 0$ in $B_1 \setminus \Lambda$,*
(iii) *$h \geq -\varepsilon_0$ in B_1,*
(iv) *$h \geq \delta_0$ in $B_1 \cap \{|x_n| \geq c_0\}$, $c_0 = 1/(16\sqrt{n})$,*

then $h \geq 0$ on $B_{1/2}$.

Proof. It is enough to show that $h \geq 0$ on $B_{1/2} \cap \{|x_n| \leq c_0\}$. Assuming the contrary, let $x^0 \in B_{1/2} \cap \{|x_n| \leq c_0\}$ be such that $h(x^0) < 0$. Consider then the auxiliary function

$$w(x) = h(x) + \frac{\alpha_0}{n-1} |x' - (x^0)'|^2 - \alpha_0 x_n^2,$$

where $\alpha_0 = \delta_0 / c_0^2$. It is immediate to check that

$$w(x^0) < 0, \quad \Delta w \leq 0 \quad \text{in } B_1 \setminus \Lambda.$$

Consider now the function w in the set $U = B_{3/4} \cap \{|x_n| \leq c_0\} \setminus \Lambda$. By the maximum principle, we must have

$$\inf_{\partial U} w < 0.$$

Analyzing the different parts of ∂U we show that the above inequality cannot hold:

1) On $\Lambda \cap U$ we have $w \geq 0$.

2) On $\{|x_n| = c_0\} \cap \partial U$ we have

$$w(x) \geq h(x) - \alpha_0 x_n^2 \geq \delta_0 - \alpha_0 c_0^2 \geq 0.$$

3) On $\{|x| = 3/4\} \cap \partial U$ we have

$$\begin{aligned} w(x) &\geq -\varepsilon_0 + \frac{\alpha_0}{n-1}|x' - (x^0)'|^2 - \alpha_0 x_n^2 \\ &\geq -\varepsilon_0 + \frac{\alpha_0}{(n-1)}\left[\frac{1}{4} - c_0\right]^2 - \alpha_0 c_0^2 \geq -\varepsilon_0 + \alpha_0\lambda_0, \end{aligned}$$

where

$$\lambda_0 = \frac{1}{(n-1)}\left[\frac{1}{4} - c_0\right]^2 - c_0^2 \geq \frac{1}{64(n-1)} - \frac{1}{256n} > 0.$$

Now, if $\varepsilon_0 < \alpha_0\lambda_0$, then $w \geq 0$ on this portion of ∂U.

Thus, we arrive at a contradiction with the assumption that $h(x^0) < 0$, and the proof is complete. □

Proof of Proposition 9.18. Assume that for a subsequence $r = r_j \to 0+$, the rescalings u_{r_j} converge to a homogeneous global solution

$$u_0(x) = C_0 \operatorname{Re}(x_{n-1} + i|x_n|)^{3/2}.$$

Then by the calculations preceding Lemma 9.20 and by the uniform convergence of $\partial_e u_{r_j}$ to $\partial_e u_0$, we obtain that for sufficiently large $j \geq j_\eta$,

$$\begin{aligned} \partial_e u_{r_j}(x) &\geq -\varepsilon_0, \quad x \in B_1, \\ \partial_e u_{r_j}(x', x_n) &\geq 2\delta_0 > 0, \quad x \in B_1 \cap \{|x_n| \geq c_0\}, \end{aligned}$$

for any $e \in \mathcal{C}'_\eta$, where c_0 and δ_0 are as in Lemma 9.20. Since also $\partial_e u_{r_j}$ is harmonic off $B_1 \setminus \Lambda(u_{r_j})$ and vanishes on $\Lambda(u_{r_j})$, we may invoke Lemma 9.20 to conclude that

$$\partial_e u_{r_j} \geq 0 \quad \text{in } B_{1/2} \quad \text{for any } e \in \mathcal{C}'_\eta,$$

for $j \geq j_\eta$. Scaling back we obtain

$$\partial_e u \geq 0 \quad \text{in } B_{r_\eta} \quad \text{for any } e \in \mathcal{C}'_\eta,$$

where $r_\eta = r_{j_\eta}/2$. Now a standard argument (see Exercise 4.1) implies that

$$\{u(x', 0) > 0\} \cap B'_{r_\eta} = \{x_{n-1} > g(x'') : x'' \in B''_{r_\eta}\},$$

where $x'' \mapsto g(x'')$ is a Lipschitz continuous function satisfying

$$|\nabla_{x''} g| \leq \eta.$$

This completes the proof of the proposition. □

Proof of Proposition 9.19. We know that for a given $\eta > 0$, $\partial_e u \geq 0$ in a small ball B_{r_η} for any tangential direction $e \in \mathcal{C}'_\eta$. Besides, the $\partial_e u$ are harmonic in $B_{r_\eta} \setminus \Lambda(u)$, and we would like to apply the boundary Harnack principle in Theorem 6.8, and to conclude, as in the proof of Theorem 6.9, that

$$\frac{\partial_{e_i} u}{\partial_{e_{n-1}} u} \in C^\alpha(B_\rho), \quad i = 1, \dots, n-2.$$

The problem, however, is that the domain $B_\rho \setminus \Lambda(u)$ is not exactly Lipschitz ($\Lambda(r)$ has a lower dimension). Nevertheless, it can be transformed into one by a bi-Lipschitz transformation. Indeed, let $z = T(x)$ be the composition of two (bi-Lipschitz) mappings $x \mapsto y \mapsto z$, given by

$$y'' = x'', \quad y_{n-1} = x_{n-1} - g(x''), \quad y_n = x_n,$$

$$z'' = y'', \quad z_{n-1} + iz_n = (y_{n-1} + iy_n)^{1/2}(y_{n-1}^2 + y_n^2)^{1/4}.$$

Note that, using polar coordinates in the (y_{n-1}, y_n)-plane we may write the last part of the transformation $y \mapsto z$ as follows: if $y_{n-1} + iy_n = \rho e^{i\theta}$, $-\pi < \theta < \pi$, then $z_{n-1} + iz_n = \rho e^{i\theta/2}$. Note that T maps $B_\rho \setminus \Lambda(u)$ into an open subset of $\{z \in \mathbb{R}^n : z_{n-1} > 0\}$ containing a half-ball $B_\delta \cap \{z_{n-1} > 0\}$, and that for any $x^j \to x^0 \in \Lambda(u)$ we can find a subsequence such that $z^j = T(x^j) \to z^0 \in \{z_{n-1} = 0\}$.

In particular, the functions $v_e(z) = \partial_e u(x(z))$ for any $e \in \mathcal{C}'_\eta$ will solve the same uniformly elliptic equation in the divergence form,

$$\operatorname{div}(A(z)\nabla v_e) = 0 \quad \text{in } B_\delta \cap \{z_{n-1} > 0\},$$

where $A(z)$ is a symmetric matrix, which is uniformly elliptic, but only measurable in z. Besides, we also know that

$$v_e = 0 \quad \text{on } B_\delta \cap \{z_{n-1} = 0\}$$

and that $v_e \geq 0$ in $B_\delta \cap \{z_{n-1} > 0\}$. Thus, by a version of the boundary Harnack principle (see remark 1) after Theorem 6.8), we obtain that

$$\frac{v_{e^1}}{v_{e^2}} \in C^\alpha(B_{\delta/2} \cap \{z_{n-1} > 0\})$$

for some $\alpha > 0$ for any $e^1, e^2 \in \mathcal{C}'_\eta$. Applying the inverse transformation $x = T^{-1}(z)$ we conclude that

$$\frac{\partial_{e^1} u}{\partial_{e^2} u} \in C^\alpha(B_\rho).$$

The rest of the proof is a repetition of the arguments in the proof of Theorem 6.9. □

9.7. The singular set

After considering $\Gamma_{3/2}(u)$, we want to study the sets $\Gamma_\kappa(u)$ with $\kappa \geq 2$. It turns out that when $\kappa = 2m$, $m \in \mathbb{N}$, the points on $\Gamma_\kappa(u)$ can be uniquely characterized as points with zero H^{n-1}-density of the coincidence set $\Lambda(u)$.

This makes them similar to the singular points in the classical obstacle problem.

Definition 9.21 (Singular points). Let u be a solution of Problem **S**. We say that $x^0 \in \Gamma(u)$ is *singular* if

$$\lim_{r\to 0+} \frac{H^{n-1}(\Lambda(u)\cap B'_r(x^0))}{H^{n-1}(B'_r)} = 0.$$

We denote the set of all singular points by $\Sigma(u)$ and call it the *singular set.*

The following proposition gives a complete characterization of singular points in terms of the value $\kappa = N(0+, u)$. The key fact used here is Proposition 9.11.

Proposition 9.22 (Characterization of singular points). *Let u be a solution of Problem* **S** *and $x^0 \in \Gamma_\kappa(u)$. Then the following statements are equivalent:*

(i) $x^0 \in \Sigma(u)$;

(ii) *any blowup of u at x^0 is a nonzero polynomial from the class*

$$\mathcal{Q}_\kappa = \{q : q \text{ is homogeneous polynomial of degree } \kappa,$$
$$\Delta q = 0,\ q(x', 0) \geq 0,\ q(x', x_n) = q(x', -x_n)\};$$

(iii) $\kappa = 2m$ *for some* $m \in \mathbb{N}$.

Proof. (i) $\Rightarrow$ (ii) The rescalings u_r satisfy

$$\Delta u_r = 2(\partial_{x_n} u_r) H^{n-1}\big|_{\Lambda(u_r)} \quad \text{in } B_1,$$

in the sense of distributions. Note that from the definition of the singular set we have $H^{n-1}(\Lambda(u_r)\cap B_1) \to 0$. Further, note that $|\nabla u_r|$ are locally uniformly bounded in B_1 by (9.11). Consequently, Δu_r converges weakly (in the sense of distributions) to 0 in B_1, and therefore any blowup u_0 must be harmonic in B_1. On the other hand, by Proposition 9.5, the function u_0 is homogeneous in B_1 and therefore can be extended by homogeneity to $\mathbb{R}^n$. The resulting extension will be harmonic in $\mathbb{R}^n$ and, being homogeneous, will have at most a polynomial growth at infinity. Then by the Liouville theorem we conclude that u_0 must be a homogeneous harmonic polynomial q of a certain integer degree κ. It is also clear that $q(x', 0) \geq 0$ and that $q(x', x_n) = q(x', -x_n)$ from the respective properties of u. Hence $q \in \mathcal{Q}_\kappa$. Finally, $q \not\equiv 0$ in $\mathbb{R}^n$ by Proposition 9.5.

(ii) $\Rightarrow$ (iii) Let q be a blowup of u at x^0. If κ is odd, the nonnegativity of q on Π implies that q vanishes on Π identically. On the other hand, from the even symmetry we also have that $\partial_{x_n} q \equiv 0$ on Π. Since q is harmonic in $\mathbb{R}^n$, the Cauchy-Kovalevskaya theorem implies that $q \equiv 0$ in $\mathbb{R}^n$, contrary to the assumption. Thus, κ is an even integer.

(iii) $\Rightarrow$ (ii) The proof is an immediate corollary of Proposition 9.11.

(ii) $\Rightarrow$ (i) Suppose that 0 is not a singular point and that over some sequence $r = r_j \to 0+$ we have $H^{n-1}(\Lambda(u_r) \cap B_1') \geq \delta > 0$. Taking a subsequence if necessary, we may assume that u_{r_j} converges to a blowup u_0. We claim that

$$H^{n-1}(\Lambda(u_0) \cap B_1') \geq \delta > 0.$$

Indeed, otherwise there exists an open set U in $\mathbb{R}^{n-1}$ with $H^{n-1}(U) < \delta$ so that $\Lambda(u_0) \cap \overline{B_1'} \subset U$. Then for large j we must have $\Lambda(u_{r_j}) \cap \overline{B_1'} \subset U$, which is a contradiction, since $H^{n-1}(\Lambda(u_{r_j}) \cap \overline{B_1'}) \geq \delta > H^{n-1}(U)$. But then u_0 vanishes identically on Π and consequently on $\mathbb{R}^n$ by the Cauchy-Kovalevskaya theorem. This completes the proof of the theorem. □

Definition 9.23. From now on we will denote

$$\Sigma_\kappa(u) := \Gamma_\kappa(u), \quad \kappa = 2m, \ m \in \mathbb{N},$$

to emphasize that the points in these sets are singular.

9.8. Weiss- and Monneau-type monotonicity formulas

In this section we introduce two one-parameter families of monotonicity formulas that we will use to study the singular set.

The first family of formulas is intrinsically related to Almgren's frequency formula, in a sense that their monotonicity follows from the same differentiation formulas for the quantities $H(r)$ and $D(r)$; see the proof of Theorem 9.4.

Theorem 9.24 (Weiss-type monotonicity formula). *Let u be a solution of Problem **S** in B_R^+ and let $\kappa \geq 0$ be any number; consider*

$$W_\kappa(r,u) := \frac{1}{r^{n-2+2\kappa}} \int_{B_r} |\nabla u|^2 - \frac{\kappa}{r^{n-1+2\kappa}} \int_{\partial B_r} u^2.$$

Then for any $0 < r < R$ one has

$$\frac{d}{dr} W_\kappa(r,u) = \frac{2}{r^{n+2\kappa}} \int_{\partial B_r} (x \cdot \nabla u - \kappa\, u)^2,$$

and consequently, $r \mapsto W_\kappa(r,u)$ is nondecreasing on $(0,1)$. Furthermore, $W_\kappa(\cdot,u)$ is constant if and only if u is homogeneous of degree κ.

We explicitly remark here that it is not necessary to assume that $0 \in \Gamma_\kappa(u)$ in the theorem, but it will be most useful in that case.

Proof. The assertion follows immediately from formulas (9.5)–(9.7), and the proof is left to the reader (see Exercise 9.7). □

With the help of Weiss-type monotonicity formulas, we can now consider a different type of blowups, which is more similar to the one in the classical obstacle case. Namely for $\kappa \geq 3/2$ and $x^0 \in \Gamma_\kappa(u)$ consider the *κ-homogeneous rescalings*

$$u^{(\kappa)}_{x^0,r}(x) = \frac{u(x^0 + rx)}{r^\kappa}.$$

Then from the growth estimate in Lemma 9.14, we obtain that the family $\{u^{(\kappa)}_{x^0,r}\}_r$ is locally uniformly bounded, and from the $C^{1,\alpha}$ estimates we may assume that $u^{(\kappa)}_{x^0,r}(x) \to u^{(\kappa)}_0(x)$ in $C^{1,\alpha}_{\mathrm{loc}}(\mathbb{R}^n_\pm \cup \Pi)$ over a certain subsequence $r = r_j \to 0+$. We call such $u^{(\kappa)}_0$ a *κ-homogeneous blowup*, in contrast to the blowups that were based on the scaling (9.8).

Proposition 9.25 (κ-homogeneous blowup). *Every blowup $u^{(\kappa)}_0$ described above is a homogeneous global solution of Problem* **S** *of degree κ.*

Note that it is a priori not obvious if $u^{(\kappa)}_0$ is nonzero.

Proof. Without loss of generality we consider the blowup at the origin. Then noticing the scaling property

$$W_\kappa(\rho, u^{(\kappa)}_r) = W_\kappa(\rho r, u),$$

where the $u^{(\kappa)}_r$ denote the κ-homogeneous rescalings at the origin, we complete the proof by a direct application of Theorem 9.24 (Exercise 9.8). □

The next family of monotonicity formulas is geared towards the study of singular points and is related to Monneau's monotonicity formula (see Theorem 7.4). Recall that by $\mathcal{Q}_\kappa$ we denote the family κ-homogeneous harmonic polynomials as in Proposition 9.22.

Theorem 9.26 (Monneau-type monotonicity formula). *Let u be a solution of Problem* **S** *in B^+_R with $0 \in \Sigma_\kappa(u)$, $\kappa = 2m$, $m \in \mathbb{N}$. Then for arbitrary $q \in \mathcal{Q}_\kappa$,*

$$r \mapsto M_\kappa(r, u, q) := \frac{1}{r^{n-1+2\kappa}} \int_{\partial B_r} (u - q)^2$$

is nondecreasing for $0 < r < R$.

Proof. The proof is completely analogous to that of Theorem 7.4 and is left to the reader (see Exercise 9.9). In fact, one can show that

$$\frac{d}{dr} M_\kappa(r, u, q) \geq \frac{2}{r} W_\kappa(r, u) \geq 0.$$

We will restrict ourselves to a remark that the computations are reduced to verifying that

$$w\Delta w = (u - q)(\Delta u - \Delta q) = -q\Delta u \geq 0$$

in the sense of measures, which indeed holds since $\Delta u \le 0$ in $\mathbb{R}^n$ and $q \ge 0$ on Π. □

Proposition 9.27 (Uniqueness of the homogeneous blowup at singular points). *Let u be a solution of Problem* **S** *and $x^0 \in \Sigma_\kappa(u)$, $\kappa = 2m$, $m \in \mathbb{N}$. Then there exists a unique nonzero $q_{x^0}^{(\kappa)} \in \mathcal{Q}_\kappa$ such that*

$$u_{x^0,r}^{(\kappa)}(x) = \frac{u(x^0 + rx)}{r^\kappa} \to q_{x^0}^{(\kappa)}(x) \quad \textit{in } C^{1,\alpha}_{\mathrm{loc}}(\mathbb{R}^n_\pm \cup \Pi).$$

Proof. Let q be a κ-homogeneous blowup over a certain sequence $r = r_j \to 0+$. Then by Propositions 9.25 and 9.11 we know that $q \in \mathcal{Q}_\kappa$.

Next, to show the uniqueness we simply repeat the arguments in the proof of Corollary 7.6, by using the monotonicity of $M_\kappa(r, u, q)$. Thus, the only thing that is left to show is that $q \neq 0$. This is rather nontrivial, and we prefer to state it as a separate lemma (see below). □

Lemma 9.28 (Nondegeneracy at singular points). *Let u be a solution of Problem* **S** *in B_1 and $0 \in \Sigma_\kappa(u)$. There exists $c > 0$, possibly depending on u, such that*

$$\sup_{\partial B_r} |u(x)| \ge c\, r^\kappa \quad \textit{for } \ 0 < r < 1.$$

Proof. Assume the contrary. Then for a sequence $r = r_j \to 0$ one has

$$h_r := \left(\frac{1}{r^{n-1}} \int_{\partial B_r} u^2 \right)^{1/2} = o(r^\kappa).$$

Passing to a subsequence if necessary, we may assume that

$$u_r(x) = \frac{u(rx)}{h_r} \to p(x) \quad \text{uniformly on } \partial B_1,$$

for some $p \in \mathcal{Q}_\kappa$; see Proposition 9.22. Note that p is nonzero as it must satisfy $\int_{\partial B_1} p^2 = 1$. Now consider the functional $M_\kappa(r, u, p)$. From the assumption on the growth of u it is easy to see that

$$M_\kappa(0+, u, p) = \int_{\partial B_1} p^2 = \frac{1}{r^{n-1+2\kappa}} \int_{\partial B_r} p^2.$$

Hence, we have that

$$\frac{1}{r^{n-1+2\kappa}} \int_{\partial B_r} (u - p)^2 \ge \frac{1}{r^{n-1+2\kappa}} \int_{\partial B_r} p^2,$$

or equivalently,

$$\int_{\partial B_r} u^2 - 2up \ge 0.$$

On the other hand, rescaling, we obtain

$$\int_{\partial B_1} h_r^2 u_r^2 - 2h_r r^\kappa u_r p \geq 0.$$

Factoring out $h_r r^\kappa$, we have

$$\int_{\partial B_1} \frac{h_r}{r^\kappa} u_r^2 - 2u_r p \geq 0,$$

and passing to the limit over $r = r_j \to 0$,

$$-\int_{\partial B_1} p^2 \geq 0.$$

Since $p \neq 0$, we have thus reached a contradiction. □

One can in fact prove a more refined version of Proposition 9.27.

Proposition 9.29. *Let u be a solution of Problem* **S**. *For $x^0 \in \Sigma_\kappa(u)$, $\kappa = 2m$, $m \in \mathbb{N}$, denote by $q_{x^0}^{(\kappa)}$ the κ-homogeneous blowup of u at x^0 as in Proposition 9.27. Then the mapping $x^0 \mapsto q_{x^0}^{(\kappa)}$ from $\Sigma_\kappa(u)$ to $\mathcal{Q}_\kappa$ is continuous. Moreover, for any compact $K \Subset \Sigma_\kappa(u)$ there exists a modulus of continuity $\sigma(r)$, such that*

$$|u(x) - q_{x^0}^{(\kappa)}(x - x^0)| \leq \sigma(|x - x^0|)|x - x^0|^\kappa,$$

for any $x_0 \in K$, $x \in D$.

Proof. Note that since $\mathcal{Q}_\kappa$ is a convex subset of a finite-dimensional vector space, namely the space of all κ-homogeneous polynomials, all the norms on such a space are equivalent. We can then endow $\mathcal{Q}_\kappa$ with the norm of $L^2(\partial B_1)$.

Now, for $x^0 \in \Sigma_\kappa(u)$ and $\varepsilon > 0$, fix $r_\varepsilon = r_\varepsilon(x_0)$ such that

$$M_\kappa(r_\varepsilon, u, q_{x^0}^{(\kappa)}, x^0) = \frac{1}{r_\varepsilon^{n-1+2\kappa}} \int_{\partial B_{r_\varepsilon}} \left(u(x + x^0) - q_{x^0}^{(\kappa)}\right)^2 < \varepsilon.$$

There exists $\delta_\varepsilon = \delta_\varepsilon(x^0)$ such that if $x^1 \in \Sigma_\kappa(u)$ and $|x^1 - x^0| < \delta_\varepsilon$, then

$$M_\kappa(r_\varepsilon, u, q_{x^0}^{(\kappa)}, x^1) = \frac{1}{r_\varepsilon^{n-1+2\kappa}} \int_{\partial B_{r_\varepsilon}} \left(u(x + x^1) - q_{x^0}^{(\kappa)}\right)^2 < 2\varepsilon.$$

From the monotonicity of Monneau's functional, we will have that

$$M_\kappa(r, u, q_{x^0}^{(\kappa)}, x^1) < 2\varepsilon, \quad 0 < r < r_\varepsilon.$$

Letting $r \to 0$, we will therefore obtain

$$M(0+, u, q_{x^0}^{(\kappa)}, x^1) = \int_{\partial B_1} \left(q_{x^1}^{(\kappa)} - q_{x^0}^{(\kappa)}\right)^2 \leq 2\varepsilon.$$

This proves the first part of the proposition.

To prove the second part, we note that

$$\begin{aligned}\|u(\cdot+x^1)-q^{(\kappa)}_{x^1}\|_{L^2(\partial B_r)} &\leq \|u(\cdot+x^1)-q^{(\kappa)}_{x^0}\|_{L^2(\partial B_r)}+\|q^{(\kappa)}_{x^1}-q^{(\kappa)}_{x^0}\|_{L^2(\partial B_r)}\\ &\leq 2(2\varepsilon)^{\frac12}r^{\frac{n-1}{2}+\kappa},\end{aligned}$$

for $|x^1-x^0|<\delta_\varepsilon$, $0<r<r_\varepsilon$, or equivalently

$$\|u^{(\kappa)}_{x^1,r}-q^{(\kappa)}_{x^1}\|_{L^2(\partial B_1)}\leq 2(2\varepsilon)^{\frac12}. \tag{9.13}$$

Now, covering the compact $K\subset\Sigma_\kappa(u)$ with finitely many balls $B_{\delta_\varepsilon(y^i)}(y^i)$, for some $y^i\in K$, $i=1,\dots,N$, we conclude that (9.13) is satisfied for all $x^1\in K$ with $r<r^K_\varepsilon:=\min\{r_\varepsilon(y^i):i=1,\dots,N\}$. On the other hand, both $u^{(\kappa)}_{x^1,r}$ and $q^{(\kappa)}_{x^1}$ are solutions of Problem **S** in B_1. Then by Exercise 9.5,

$$\left(u^{(\kappa)}_{x^1,r}-q^{(\kappa)}_{x^1}\right)^{\pm}$$

are subharmonic, and therefore using L^∞-L^2 estimates for subharmonic functions, we obtain

$$\|u^{(\kappa)}_{x^1,r}-q^{(\kappa)}_{x^1}\|_{L^\infty(B_{1/2})}\leq C_{n,\varepsilon}$$

for $x^1\in K$, $0<r<r^K_\varepsilon$ with $C_{n,\varepsilon}\to 0$ as $\varepsilon\to 0$. Clearly, this implies the second part of the theorem. □

9.9. The structure of the singular set

In this section we will prove a structural theorem for the singular set $\Sigma_\kappa(u)$, similar to the one for the classical obstacle problem (see Theorem 7.9).

For a singular point $x^0\in\Sigma_\kappa(u)$, $\kappa=2m$, $m\in\mathbb{N}$, we consider the unique κ-homogeneous blowup $q^{(\kappa)}_{x^0}$ as in Propositions 9.27 and define the "tangent plane"

$$\Pi^{(\kappa)}_{x^0}:=\{\xi\in\mathbb{R}^{n-1}:\xi\cdot\nabla_{x'}q^{(\kappa)}_{x^0}(x',0)\equiv 0 \text{ on } \mathbb{R}^{n-1}\}$$

and its dimension

$$d^{(\kappa)}_{x^0}:=\dim\Pi^{(\kappa)}_{x^0}.$$

Note that since $q^{(\kappa)}_{x^0}\not\equiv 0$ on $\mathbb{R}^n$ (see Proposition 9.27), we then have $q^{(\kappa)}_{x^0}\not\equiv 0$ on Π (by Exercise 9.10) and therefore

$$d^{(\kappa)}_{x^0}\in\{0,\dots,n-2\}.$$

Theorem 9.30 (Structure of the singular set). *Let u be a solution of Problem* **S**. *Then for every $\kappa=2m$, $m\in\mathbb{N}$, and $d=0,1,\dots,n-2$, the set*

$$\Sigma^d_\kappa(u):=\{x^0\in\Sigma_\kappa(u):d^{(\kappa)}_{x^0}=d\}$$

is contained in a countable union of d-dimensional C^1 manifolds.

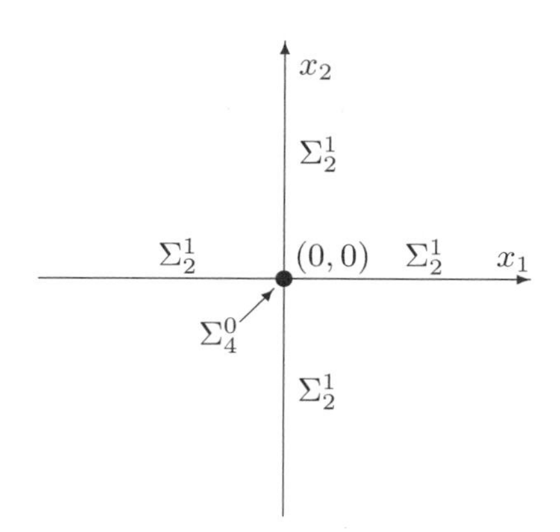

Figure 9.3. Free boundary for $q(x) = x_1^2x_2^2 - (x_1^2 + x_2^2)\,x_3^2 + \frac{1}{3}x_3^4$ in $\mathbb{R}^3$ with zero thin obstacle on $\mathbb{R}^2 \times \{0\}$

The following example provides an illustration of Theorem 9.30. Consider the harmonic polynomial $q(x) = x_1^2x_2^2 - (x_1^2 + x_2^2)\,x_3^2 + \frac{1}{3}x_3^4$ in $\mathbb{R}^3$. Note that $q \in \mathfrak{Q}_4$. On $\Pi = \mathbb{R}^2 \times \{0\}$, we have $u(x_1, x_2, 0) = x_1^2x_2^2$, and therefore the coincidence set $\Lambda(q)$ as well as the free boundary $\Gamma(q)$ consists of the union of the lines $\mathbb{R} \times \{0\} \times \{0\}$ and $\{0\} \times \mathbb{R} \times \{0\}$. Thus, all free boundary points are singular. It is straightforward to check that $0 \in \Sigma_4^0(q)$ and that the rest of the free boundary points are in $\Sigma_2^1(q)$; see Figure 9.3.

The proof of Theorem 9.30 is based on Whitney's extension theorem; see Lemma 7.10. An essential condition for the existence of the extension is the closedness of the set E. However, notice that the singular set $\Sigma_\kappa(u)$ is not generally closed; see the example above. Nevertheless the nondegeneracy at singular points (Lemma 9.28) implies that the $\Sigma_\kappa(u)$ are of type F_σ.

Lemma 9.31 ($\Sigma_\kappa(u)$ is F_σ). *For any solution u of Problem* **S**, *the set $\Sigma_\kappa(u)$ is of type F_σ, i.e., it is a union of countably many closed sets.*

Proof. Without loss of generality, consider the solutions in B_1^+. Let E_j be the set of points $x^0 \in \Sigma_\kappa(u) \cap \overline{B_{1-1/j}}$ such that

$$\frac{1}{j}\,\rho^\kappa \leq \sup_{|x-x_0|=\rho} |u(x)| < j\rho^\kappa, \tag{9.14}$$

for $0 < \rho < 1 - |x_0|$. Note that from Lemmas 9.14 and 9.28 we have that

$$\Sigma_\kappa(u) = \bigcup_{j=1}^\infty E_j.$$

The lemma will follow once we show that E_j is a closed set. Indeed, if $x_0 \in \overline{E}_j$, then x_0 satisfies (9.14) and we need to show only that $x^0 \in \Sigma_\kappa(u)$, or equivalently that $N(0+, u, x^0) = \kappa$. Since the function $x \mapsto N(0+, u, x)$ is upper semicontinuous, we readily have that $N(0+, u, x^0) \geq \kappa$. On the other hand, if $N(0+, u, x^0) = \kappa' > \kappa$, then by Lemma 9.14 we would have

$$|u(x)| < C|x - x^0|^{\kappa'} \quad \text{in } B_{1-|x_0|}(x_0),$$

which would contradict the lower estimate in (9.14). Thus $N(0+, u, x^0) = \kappa$ and therefore $x^0 \in E_j$. □

Proof of Theorem 9.30. Without loss of generality, we assume that $R = 1$.

Step 1: Whitney's extension. Let $K = E_j$ be the compact subset of $\Sigma_\kappa(u)$ defined in the proof of Lemma 9.31. Next, define the family of polynomials

$$p_{x^0}(x) = q_{x^0}^{(\kappa)}(x - x^0);$$

we claim that they satisfy the conditions of Whitney's extension theorem on the compact set E_j. We have to verify the compatibility condition (ii) in Lemma 7.10 up to the order κ, i.e.,

$$|\partial^\alpha p_{x^0}(x^1) - \partial^\alpha p_{x^1}(x^1)| = o(|x^0 - x^1|^{\kappa-|\alpha|}) \quad \text{for any } x^0, x^1 \in E_j, \tag{9.15}$$

for any multiindex α of order $|\alpha| \leq \kappa$.

1) Consider first the case $|\alpha| = \kappa$. Note that

$$\partial^\alpha p_{x^0}(x) = a_\alpha(x^0)$$

is the Taylor coefficient of the polynomial p_{x^0} at x^0 so that

$$p_{x^0}(x) = \sum_{|\alpha|=\kappa} a_\alpha(x^0) \frac{(x - x^0)^\alpha}{\alpha!}.$$

Then condition (9.15) for $|\alpha| = \kappa$ becomes

$$a_\alpha(x^0) - a_\alpha(x^1) = o(x^1 - x^0), \quad x^0, x^1 \in E_j,$$

which is equivalent to the continuity of the mapping $x^0 \mapsto q_{x^0}^{(\kappa)}$ on E_j. This is established in Proposition 9.29.

2) For $0 \leq |\alpha| < \kappa$ we have $\partial^\alpha p_{x^1}(x^1) = 0$ and therefore the compatibility condition (9.15) becomes

$$|\partial^\alpha p_{x^0}(x^1)| \leq \sigma(|x^1 - x^0|)|x^1 - x^0|^{\kappa-|\alpha|}, \quad x^0, x^1 \in E_j,$$

for a certain modulus of continuity $\sigma(r)$. We can explicitly write that

$$\partial^\alpha p_{x^0}(x) = \partial^\alpha q_{x^0}(x - x^0) = \sum_{\gamma>\alpha, |\gamma|=\kappa} \frac{a_\gamma(x^0)}{(\gamma - \alpha)!}(x - x^0)^{\gamma-\alpha},$$

and therefore if (9.15) is violated, then we can find $\delta > 0$ and sequences $x^i, y^i \in E_j$ with

$$|y^i - x^i| =: \rho_i \to 0,$$

such that

$$(9.16)\qquad \Big| \sum_{\gamma>\alpha,|\gamma|=\kappa} \frac{a_\gamma(x^i)}{(\gamma-\alpha)!}(y^i - x^i)^{\gamma-\alpha}\Big| \geq \delta |y^i - x^i|^{\kappa-|\alpha|}.$$

Consider the rescalings

$$w^i(x) = \frac{u(x^i + \rho_i x)}{\rho_i^\kappa}, \quad \xi^i = (y^i - x^i)/\rho_i.$$

Without loss of generality we may assume that $x^i \to x^0 \in E_j$ and $\xi^i \to \xi^0 \in \partial B_1$. From Proposition 9.29 we have that

$$|w^i(x) - q_{x^i}^{(\kappa)}(x)| \leq \sigma(\rho_i|x|)|x|^\kappa,$$

and therefore $w^i(x)$ converges locally uniformly in $\mathbb{R}^n$ to $q_{x^0}^{(\kappa)}(x)$. Further, note that since y^i and x^i are from the set E_j, the inequalities (9.14) are satisfied there. Moreover, we also have that similar inequalities are satisfied for the rescaled function w^i at 0 and ξ^i. Therefore, passing to the limit, we obtain

$$\tfrac{1}{j}\rho^\kappa \leq \sup_{|x-\xi^0|=\rho} q_{x^0}^{(\kappa)}(x) \leq j\rho^\kappa, \quad 0 < \rho < \infty.$$

This implies that ξ^0 is a point of frequency $\kappa = 2m$ for the polynomial $q_{x^0}^{(\kappa)}$, i.e., $\xi^0 \in \Sigma_\kappa(q_{x^0}^{(\kappa)})$. In particular,

$$\partial^\alpha q_{x^0}^{(\kappa)}(\xi^0) = 0, \quad \text{for } |\alpha| < \kappa.$$

However, dividing both parts of (9.16) by $\rho_i^{\kappa-|\alpha|}$ and passing to the limit, we obtain that

$$|\partial^\alpha q_{x^0}^{(\kappa)}(\xi_0)| = \Big| \sum_{\gamma>\alpha,|\gamma|=\kappa} \frac{a_\gamma(x_0)}{(\gamma-\alpha)!}(\xi_0)^{\gamma-\alpha}\Big| \geq \delta,$$

a contradiction.

So in all cases, the compatibility conditions (9.15) are satisfied and we can apply Whitney's extension theorem. Thus, there exists a function $F \in C^\kappa(\mathbb{R}^n)$ such that

$$\partial^\alpha F(x^0) = \partial^\alpha p_{x^0}(x^0) = \partial^\alpha q_{x^0}^{(\kappa)}(0) \quad \text{on } E_j$$

for any $|\alpha| \leq \kappa$.

Step 2: Implicit function theorem. Suppose now that $x^0 \in \Sigma^d_\kappa(u) \cap E_j$. Recalling the definition of $d^{(\kappa)}_{x^0}$, this means that

$$d = \dim\{\xi \in \mathbb{R}^{n-1} : \xi \cdot \nabla_{x'} q^{(\kappa)}_{x^0}(x', 0) \equiv 0\}.$$

This is equivalent to saying that there exist multiindices β_i of order $|\beta_i| = \kappa - 1$, $i = 1, 2, \dots, n-1-d$, such that the vectors

$$\nabla_{x'} \partial^{\beta_i} q^{(\kappa)}_{x^0}(0)$$

are linearly independent.

This can be written as

$$\det(\partial_{\nu_k} \partial^{\beta^i} F(x_0)) \neq 0, \quad i, k = 1, \dots, n-1-d, \tag{9.17}$$

for some directions ν_k. On the other hand,

$$\Sigma^d_\kappa(u) \cap E_j \subset \bigcap_{i=1}^{n-1-d} \{\partial^{\beta^i} F = 0\}.$$

Therefore, in view of the implicit function theorem, condition (9.17) implies that $\Sigma^d_\kappa(u) \cap E_j$ is contained in a d-dimensional manifold in a neighborhood of x^0. Finally, since $\Sigma_k(u) = \bigcup_{j=1}^\infty E_j$, this implies the statement of the theorem. □

Notes

From the earlier works on the regularity of solutions we mention those by Frehse [**Fre75, Fre77**], who established the Lipschitz regularity as well as the C^1 regularity along tangential directions. The $C^{1,1/2}$ (optimal) regularity in dimension 2 was proved by Richardson [**Ric78**]. The $C^{1,\alpha}$ regularity in arbitrary dimensions was proved by Caffarelli [**Caf79**]. Our proof of $C^{1,\alpha}$ regularity is based on a more robust approach of Uraltseva [**Ura85**], which allows easier generalization to the time-dependent case. The $C^{1,1/2}$ regularity in arbitrary dimensions was first proved by Athanasopoulos-Caffarelli [**AC04**]. Our proof is based on a different approach (using Almgren's frequency formula) taken by Caffarelli-Salsa-Silvestre [**CSS08**], which also works for the obstacle problem for arbitrary fractional powers of the Laplacian. Almgren's frequency formula originated in the study of multivalued harmonic functions [**Alm00**]. It was also used in the context of unique continuation by Garofalo-Lin [**GL86, GL87**]. The paper [**CSS08**] contains a new powerful "truncated" version of Almgren's frequency formula that makes it possible to treat problems with nonzero thin obstacles. It was further refined in Garofalo-Petrosyan [**GP09**]. The $C^{1,1/2}$ regularity when the hypersurface $\mathcal{M}$ (supporting the thin obstacle) is $C^{1,\beta}$, $\beta > 1/2$, was established by Guillen [**Gui09**].

There are not many results concerning the free boundary in the earlier works, except perhaps the result of Lewy [**Lew72**] that the coincidence set in dimension 2 must be a finite union of intervals when the thin obstacle is real-analytic. The $C^{1,\alpha}$ regularity of the free boundary near regular points was first proved by Athanasopoulos-Caffarelli-Salsa [**ACS08**]. It then was extended to the arbitrary fractional powers of the Laplacian by Caffarelli-Salsa-Silvestre [**CSS08**].

The study of the singular set (as presented here) is based on the work of Garofalo-Petrosyan [**GP09**]. Weiss- and Monneau-type monotonicity formulas in the context of the thin obstacle problem also originated in that paper.

The thin obstacle problem considered in this chapter is only a particular case of the Signorini problem (a system) in elastostatics (see e.g. Duvaut-Lions [**DL76**, Chapter III, §5.4.2]), when the displacement occurs only in normal direction. The existence of weak solutions was proved by Fichera [**Fic64**] and the $C^{1,\alpha}$ regularity of solutions by Schumann [**Sch89**]. Recently, the optimal $C^{1,1/2}$ regularity was established in the groundbreaking paper of Andersson [**And12**], who also proved the $C^{1,\alpha}$ regularity of the free boundary near regular points, matching the known results in the scalar case.

Exercises

9.1. Let $u \in C^{1,\alpha}(B_R^+ \cup B_R') \cap W^{1,2}(B_R^+)$. Then u is a solution of Problem **S** in B_R^+ iff u is harmonic in B_R^+ and satisfies the Signorini boundary conditions

$$u \geq 0, \quad -\partial_{x_n} u \geq 0, \quad u \partial_{x_n} u = 0 \quad \text{on } B_R'.$$

Prove the same statement if $u \in W^{2,2}_{\mathrm{loc}}(B_R^+) \cap W^{1,2}(B_R^+)$. The Signorini boundary conditions must be understood in a.e. sense in that case.

Hint: First prove that u is harmonic in B_R^+. Then show that u is a solution of Problem **S** iff

$$-\int_{B_R'} (v-u)\partial_{x_n} u = \int_{B_R^+} \nabla u \nabla (v-u) \geq 0,$$

for any $v \in \mathfrak{K} = \{v \in W^{1,2}(B_R^+) : v|_{B_R'} \geq 0,\ v|_{(\partial B_R)^+} = u\}$.

9.2. Let u be a solution of Problem **S** in B_1 and u^ε a solution of the penalized Problem $\mathbf{S}_\varepsilon$.

(i) Prove that uniformly in $\varepsilon \in (0,1)$,

$$\|u^\varepsilon\|_{W^{1,2}(B_1^+)} \leq C_n \|u\|_{W^{1,2}(B_1^+)}.$$

(ii) Show that $u^\varepsilon \to u$ weakly in $W^{1,2}(B_1^+)$.

Hint: (i) Plug $\eta = u^\varepsilon - u$ in (9.1) and use that $\beta_\varepsilon(u^\varepsilon)(u^\varepsilon - u) = (\beta_\varepsilon(u^\varepsilon) - \beta_\varepsilon(u))(u^\varepsilon - u) \geq 0$ on B_1'.

(ii) Plug $\eta = u^\varepsilon \zeta^2$ with a cutoff function $\zeta \in C_0^\infty(B_1)$ to obtain that

$$\int_{B_1'} \beta_\varepsilon(u^\varepsilon) u^\varepsilon \zeta^2 \leq C_n \|u\|^2_{W^{1,2}(B_1^+)}.$$

This implies that for any $\delta > 0$

$$H^{n-1}(\{u_\varepsilon(\cdot, 0) < -\delta\} \cap \{\zeta = 1\})\delta(\delta - \varepsilon^2) \leq C\varepsilon,$$

which implies that any weak limit $\tilde{u}$ of u^ε over any subsequence is nonnegative a.e. on B_1'. Further, arguing as in (i), note that

$$\int_{B_1^+} |\nabla u^\varepsilon|^2 \leq \int_{B_1^+} |\nabla u|^2,$$

which after passing to the limit will yield a similar inequality for $\tilde{u}$. From the minimality of u and the uniqueness of the minimizer with the same boundary data, it follows that $\tilde{u} = u$.

9.3. Prove Lemma 9.1.

Proof. 1) We first consider the case $k = 1, \dots, n-1$. Let $\tau = e_k$. In (9.1) instead of η plug $\eta_\tau = \partial_\tau \eta$ and integrate by parts. We then obtain

$$\int_{B_1^+} \nabla u_\tau^\varepsilon \nabla \eta = -\int_{B_1'} \beta_\varepsilon'(u^\varepsilon) u_\tau^\varepsilon \eta.$$

Now choose $\eta = u_\tau^\varepsilon \hat{G}(x - x^0)\zeta^2$, with $x^0 \in B_1'$, where $G(y) = 1/|y|^{n-2}$, $\hat{G} = \min\{G, 1/\delta^{n-2}\}$, and $\zeta \in C_0^\infty(B_1)$. This gives

$$\int_{B_1^+} |\nabla u_\tau^\varepsilon|^2 \hat{G}\zeta^2 + u_\tau^\varepsilon \nabla u_\tau^\varepsilon \nabla \hat{G}\zeta^2 + 2u_\tau^\varepsilon \nabla u_\tau^\varepsilon \hat{G}\zeta\nabla\zeta) \leq 0.$$

Writing the second term as $(1/2)\nabla(u_\tau^\varepsilon)^2 \nabla \hat{G}\zeta^2$ and integrating by parts, we obtain

$$\begin{aligned}\int_{B_1^+} |\nabla u_\tau^\varepsilon|^2 \hat{G}\zeta^2 \leq & \int_{B_1^+} (u_\tau^\varepsilon)^2 \nabla \hat{G}\zeta\nabla\zeta + 2u_\tau^\varepsilon \nabla u_\tau^\varepsilon \hat{G}\zeta\nabla\zeta \\ & - \frac{n-1}{2\delta^{n-1}} \int_{(\partial B_\delta)^+} (u_\tau^\varepsilon)^2 + \int_{B_1'} \frac{1}{2}(u_\tau^\varepsilon)^2 \hat{G}_{x_n}\zeta^2.\end{aligned}$$

Now, the last term vanishes, since $\hat{G}_{x_n} = 0$ on B_1', and therefore arguing as in the proof of the energy inequality, we obtain

$$\int_{B_1^+} |\nabla u_\tau^\varepsilon|^2 \hat{G}\zeta^2 \leq C \int_{B_1^+} (u_\tau^\varepsilon)^2 (\hat{G}|\nabla\zeta|^2 + |\nabla\hat{G}|\zeta|\nabla\zeta|).$$

Taking $\zeta \in C_0^\infty(B_{2\rho})$ such that $\zeta = 1$ on B_ρ and $|\nabla\zeta| \le C/\rho$ and letting $\delta \to 0$ we thus obtain

$$\int_{B_\rho(x^0)} \frac{|\nabla u_\tau^\varepsilon|^2}{|x-x^0|^{n-2}} \le \frac{C}{\rho^n} \int_{B_{2\rho}(x^0)\setminus B_\rho(x^0)} (u_\tau^\varepsilon)^2.$$

At this point we note that we immediately get bounds uniform in ε on $\|\partial_{x_k x_i} u^\varepsilon\|_{L^2(B^+_{3/4})}$ for $k = 1, \dots, n$, $i = 1, \dots, n-1$. Using the equation for u^ε this also provides uniform estimates for $\|\partial_{x_n x_n} u^\varepsilon\|_{L^2(B^+_{3/4})}$, which implies the uniform boundedness of the family $\{u^\varepsilon\}$ in $W^{2,2}(B^+_{3/4})$.

2) To prove the estimate for $k = n$, we start with the identity

$$\int_{B_1^+} \nabla u^\varepsilon \nabla\eta = -\int_{B_1'} u^\varepsilon_{x_n}\eta,$$

for any $\eta \in C^\infty(B_1)$. Then plug η_{x_n} instead of η, and integrate by parts. This gives

$$\int_{B_1^+} \nabla u^\varepsilon_{x_n} \nabla\eta = \int_{B_1'} (u^\varepsilon_{x_n}\eta_{x_n} - \nabla u^\varepsilon\nabla\eta) = -\int_{B_1'} \nabla_{x'}u^\varepsilon \nabla_{x'}\eta.$$

Now choose $\eta = u^\varepsilon_{x_n}\hat{G}\zeta^2$:

$$\begin{aligned}
&\int_{B_1^+} |\nabla u^\varepsilon_{x_n}|^2\hat{G}\zeta^2 + u^\varepsilon_{x_n}\nabla u^\varepsilon_{x_n}\nabla\hat{G}\zeta^2 + 2u^\varepsilon_{x_n}\nabla u^\varepsilon_{x_n}\hat{G}\zeta\nabla\zeta \\
&\quad = -\int_{B_1'} \nabla_{x'}u^\varepsilon\nabla_{x'}(u^\varepsilon_{x_n}\hat{G}\zeta^2) \\
&\quad = -\int_{B_1'} |\nabla_{x'}u^\varepsilon|^2\beta_\varepsilon'(u^\varepsilon)\hat{G}\zeta^2 - \int_{B_1'} \nabla_{x'}u^\varepsilon u^\varepsilon_{x_n}\nabla_{x'}(\hat{G}\zeta^2).
\end{aligned}$$

The first integral in the last line is clearly nonnegative. In the second integral we can pass to the strong limit (because of the uniform $W^{2,2}$ bounds on u^ε) to obtain that it converges to zero, since $\nabla_{x'}u\partial_{x_n}u = 0$ on $B'_{3/4}$ (this is an easy consequence of Exercise 9.1). Thus, passing to the limit as $\varepsilon \to 0$, we obtain

$$\int_{B_\rho(x^0)} \frac{|\nabla u_{x_n}|^2}{|x-x^0|^{n-2}} \le \frac{C}{\rho^n} \int_{B_{2\rho}(x^0)\setminus B_\rho(x^0)} u^2_{x_n}. \qquad \square$$

9.4. Let $v \in W^{1,2}(B_1)$ be such that

$$\int_{B_\rho(x^0)} \frac{|\nabla v(x)|^2}{|x-x^0|^{n-2}} \le C\rho^\alpha, \quad 0 < \alpha < 1,$$

for a.e. $x^0 \in B'_{1/2}$ and any $\rho \in (0, 1/4)$. Then

$$|v(x^0) - v(y^0)| \le C|x^0 - y^0|^{\alpha/2},$$

for any x^0, $y^0 \in B'_{1/2}$ with $|x^0 - y^0| < 1/8$.

Hint: 1) Show that

$$\int_{B_\rho(x^0)} \frac{|\nabla v(x)|}{|x-x^0|^{n-1}} \le C\rho^{\alpha/2}.$$

2) Prove that

$$\int_{B_\rho(x^0)} |v(x)-v(x^0)| \le C_n\rho^n \int_{B_\rho(x^0)} \frac{|\nabla v(x)|}{|x-x^0|^{n-1}},$$

for a.e. $x^0 \in B'_{1/2}$ and $\rho \in (0,1/4)$.

3) For points x^0, $y^0 \in B'_{1/2}$ with $|x^0-y^0| = h < 1/8$ show that

$$|v(x^0)-v(y^0)| \le \frac{1}{|B_h|}\int_{B_{2h}(x^0)} |v(z)-v(x^0)|dz + \frac{1}{|B_h|}\int_{B_{2h}(y^0)} |v(z)-v(y^0)|dz.$$

9.5. Show that if u and v are solutions of Problem **S** in B_R^+ (evenly extended to B_R), then $(u-v)^\pm$ are subharmonic functions in B_R. Moreover, if $e \in \partial B'_1$, then $(\partial_e u)^\pm$ are also subharmonic.

Hint: Use Exercise 2.6.

9.6. Show that in dimension $n=2$, the homogeneous harmonic polynomials

$$\hat{v}_\kappa(x) = C_\kappa \operatorname{Im}(x_1 + i|x_2|)^\kappa, \quad \kappa = 2m+1, \ m \in \mathbb{N}$$

can never occur as a blowup at a free boundary point, even though they are global solutions of Problem **S**.

Proof. Let u be a solution of Problem **S** with $0 \in \Gamma(u) = \partial\{u(\cdot,0)>0\}$. Then we can choose a sequence $r = r_j \to 0+$ so that $u(\frac{1}{2}r_j,0)>0$ (or $u(-\frac{1}{2}r_j,0)>0$). Then from the complementarity condition in (**S**) we will have $\partial_{x_2}u(\frac{1}{2}r_j,0)=0$ implying that $\partial_{x_2}u_{r_j}(\frac{1}{2},0)=0$. Hence, if u_0 is a blowup over a subsequence of $\{r_j\}$, the C^1 convergence will imply that $\partial_{x_2}u_0(\frac{1}{2},0)=0$. However, $\hat{v}_\kappa$ do not satisfy this condition. □

9.7. Prove Theorem 9.24.

9.8. Complete the proof of Proposition 9.25

9.9. Prove Theorem 9.26.

9.10. Prove that every polynomial $q \in \mathcal{Q}_\kappa$ can be recovered uniquely from its restriction to Π. In particular, if $q \not\equiv 0$ on $\mathbb{R}^n$ then $q \not\equiv 0$ on Π.

Hint: Use the Cauchy-Kovalevskaya theorem.

Bibliography

[AC85] Ioannis Athanasopoulos and Luis A. Caffarelli, *A theorem of real analysis and its application to free boundary problems*, Comm. Pure Appl. Math. **38** (1985), no. 5, 499–502. MR803243 (86j:49062)

[AC04] I. Athanasopoulos and L. A. Caffarelli, *Optimal regularity of lower dimensional obstacle problems*, Zap. Nauchn. Sem. S.-Peterburg. Otdel. Mat. Inst. Steklov. (POMI) **310** (2004), Kraev. Zadachi Mat. Fiz. i Smezh. Vopr. Teor. Funkts. 35 [34], 49–66, 226. MR2120184 (2006i:35053)

[ACF84] Hans Wilhelm Alt, Luis A. Caffarelli, and Avner Friedman, *Variational problems with two phases and their free boundaries*, Trans. Amer. Math. Soc. **282** (1984), no. 2, 431–461. MR732100 (85h:49014)

[ACS08] I. Athanasopoulos, L. A. Caffarelli, and S. Salsa, *The structure of the free boundary for lower dimensional obstacle problems*, Amer. J. Math. **130** (2008), no. 2, 485–498. MR2405165 (2009g:35345)

[ADN59] S. Agmon, A. Douglis, and L. Nirenberg, *Estimates near the boundary for solutions of elliptic partial differential equations satisfying general boundary conditions. I*, Comm. Pure Appl. Math. **12** (1959), 623–727. MR0125307 (23 #A2610)

[Alm00] Frederick J. Almgren, Jr., *Almgren's big regularity paper*, World Scientific Monograph Series in Mathematics, vol. 1, World Scientific Publishing Co. Inc., River Edge, NJ, 2000, Q-valued functions minimizing Dirichlet's integral and the regularity of area-minimizing rectifiable currents up to codimension 2, With a preface by Jean E. Taylor and Vladimir Scheffer. MR1777737 (2003d:49001)

[ALM03] Hiroaki Aikawa, Torbjörn Lundh, and Tomohiko Mizutani, *Martin boundary of a fractal domain*, Potential Anal. **18** (2003), no. 4, 311–357. MR1953266 (2004a:31002)

[ALS12] J. Andersson, E. Lindgren, and H. Shahgholian, *Optimal regularity for the no-sign obstacle problem*, to appear, 2012.

[Alt77] Hans Wilhelm Alt, *The fluid flow through porous media. Regularity of the free surface*, Manuscripta Math. **21** (1977), no. 3, 255–272. MR0449170 (56 #7475)

[AMM06] John Andersson, Norayr Matevosyan, and Hayk Mikayelyan, *On the tangential touch between the free and the fixed boundaries for the two-phase obstacle-like problem*, Ark. Mat. **44** (2006), no. 1, 1–15. MR2237208 (2007e:35296)

[AM12] John Andersson and Hayk Mikayelyan, *On the non-tangential touch between the free and the fixed boundaries for the two-phase obstacle-like problem*, Math. Ann. **352** (2012), no. 2, 357–372.

[AMU09] D. E. Apushkinskaya, N. Matevosyan, and N. N. Uraltseva, *The behavior of the free boundary close to a fixed boundary in a parabolic problem*, Indiana Univ. Math. J. **58** (2009), no. 2, 583–604. MR2514381 (2010g:35343)

[And07] John Andersson, *On the regularity of a free boundary near contact points with a fixed boundary*, J. Differential Equations **232** (2007), no. 1, 285–302. MR2281197 (2008e:35210)

[And12] ———, *Optimal regularity for the Signorini problem and its free boundary*, Algebra i Analiz **24** (2012), to appear.

[AS05] John Andersson and Henrik Shahgholian, *Global solutions of the obstacle problem in half-spaces, and their impact on local stability*, Calc. Var. Partial Differential Equations **23** (2005), no. 3, 271–279. MR2142064 (2006b:35352)

[ASU00] D. E. Apushkinskaya, H. Shahgholian, and N. N. Uraltseva, *Boundary estimates for solutions of a parabolic free boundary problem*, Zap. Nauchn. Sem. S.-Peterburg. Otdel. Mat. Inst. Steklov. (POMI) **271** (2000), Kraev. Zadachi Mat. Fiz. i Smezh. Vopr. Teor. Funkts. 31, 39–55, 313. MR1810607 (2002b:35205)

[ASW10] John Andersson, Henrik Shahgholian, and Georg S. Weiss, *Uniform regularity close to cross singularities in an unstable free boundary problem*, Comm. Math. Phys. **296** (2010), no. 1, 251–270. MR2606634 (2011c:35623)

[ASW12] ———, *On the singularities of a free boundary through Fourier expansion*, to appear, 2012.

[AU95] D. E. Apushkinskaya and N. N. Uraltseva, *On the behavior of the free boundary near the boundary of the domain*, Zap. Nauchn. Sem. S.-Peterburg. Otdel. Mat. Inst. Steklov. (POMI) **221** (1995), Kraev. Zadachi Mat. Fiz. i Smezh. Voprosy Teor. Funktsii. 26, 5–19, 253. MR1359745 (96m:35340)

[AU07] ———, *Boundary estimates for solutions of two-phase obstacle problems*, J. Math. Sci. (N. Y.) **142** (2007), no. 1, 1723–1732, Problems in mathematical analysis. No. 34. MR2331633 (2009c:35482)

[AU09] ———, *Boundary estimates for solutions to the two-phase parabolic obstacle problem*, J. Math. Sci. (N. Y.) **156** (2009), no. 4, 569–576, Problems in mathematical analysis. No. 38. MR2493233 (2010d:35412)

[AUS02] D. E. Apushkinskaya, N. N. Uraltseva, and H. Shahgolian, *On global solutions of a parabolic problem with an obstacle*, St. Petersburg Math. J. **14** (2003), no. 1, 1–17. MR1893318 (2003d:35163)

[AUS03] ———, *On the Lipschitz property of the free boundary in a parabolic problem with an obstacle*, St. Petersburg Math. J. **15** (2004), no. 3, 375–391. MR2052937 (2005d:35276)

[AW06] J. Andersson and G. S. Weiss, *Cross-shaped and degenerate singularities in an unstable elliptic free boundary problem*, J. Differential Equations **228** (2006), no. 2, 633–640. MR2289547 (2007k:35522)

[BBC94] H. Berestycki, A. Bonnet, and S. J. Chapman, *A semi-elliptic system arising in the theory of type-II superconductivity*, Comm. Appl. Nonlinear Anal. **1** (1994), no. 3, 1–21. MR1295490 (95e:35192)

[BK74] Haïm Brezis and David Kinderlehrer, *The smoothness of solutions to nonlinear variational inequalities*, Indiana Univ. Math. J. **23** (1973/74), 831–844. MR0361436 (50 #13881)

[Bla01] Ivan Blank, *Sharp results for the regularity and stability of the free boundary in the obstacle problem*, Indiana Univ. Math. J. **50** (2001), no. 3, 1077–1112. MR1871348 (2002i:35213)

[Bla04] ______, *Eliminating mixed asymptotics in obstacle type free boundary problems*, Comm. Partial Differential Equations **29** (2004), no. 7-8, 1167–1186. MR2097580 (2005h:35372)

[BM00] A. Bonnet and R. Monneau, *Distribution of vortices in a type-II superconductor as a free boundary problem: existence and regularity via Nash-Moser theory*, Interfaces Free Bound. **2** (2000), no. 2, 181–200. MR1760411 (2001e:35180)

[Bre72] Haïm Brezis, *Problèmes unilatéraux*, J. Math. Pures Appl. (9) **51** (1972), 1–168. MR0428137 (55 #1166)

[BS68] Haïm R. Brezis and Guido Stampacchia, *Sur la régularité de la solution d'inéquations elliptiques*, Bull. Soc. Math. France **96** (1968), 153–180. MR0239302 (39 #659)

[Caf77] Luis A. Caffarelli, *The regularity of free boundaries in higher dimensions*, Acta Math. **139** (1977), no. 3-4, 155–184. MR0454350 (56 #12601)

[Caf79] L. A. Caffarelli, *Further regularity for the Signorini problem*, Comm. Partial Differential Equations **4** (1979), no. 9, 1067–1075. MR542512 (80i:35058)

[Caf80] Luis A. Caffarelli, *Compactness methods in free boundary problems*, Comm. Partial Differential Equations **5** (1980), no. 4, 427–448. MR567780 (81e:35121)

[Caf81] ______, *A remark on the Hausdorff measure of a free boundary, and the convergence of coincidence sets*, Boll. Un. Mat. Ital. A (5) **18** (1981), no. 1, 109–113. MR607212 (82i:35078)

[Caf93] ______, *A monotonicity formula for heat functions in disjoint domains*, Boundary value problems for partial differential equations and applications, RMA Res. Notes Appl. Math., vol. 29, Masson, Paris, 1993, pp. 53–60. MR1260438 (95e:35096)

[Caf98a] L. A. Caffarelli, *The obstacle problem revisited*, J. Fourier Anal. Appl. **4** (1998), no. 4-5, 383–402. MR1658612 (2000b:49004)

[Caf98b] Luis A. Caffarelli, *The obstacle problem*, Lezioni Fermiane [Fermi Lectures], Accademia Nazionale dei Lincei, Rome, 1998. MR2011808 (2004g:49002)

[CGI$^+$00] S. Chanillo, D. Grieser, M. Imai, K. Kurata, and I. Ohnishi, *Symmetry breaking and other phenomena in the optimization of eigenvalues for composite membranes*, Comm. Math. Phys. **214** (2000), no. 2, 315–337. MR1796024 (2001i:49077)

[Cha95] S. Jonathan Chapman, *A mean-field model of superconducting vortices in three dimensions*, SIAM J. Appl. Math. **55** (1995), no. 5, 1259–1274. MR1349309 (97d:82085)

[CJK02] Luis A. Caffarelli, David Jerison, and Carlos E. Kenig, *Some new monotonicity theorems with applications to free boundary problems*, Ann. of Math. (2) **155** (2002), no. 2, 369–404. MR1906591 (2003f:35068)

[CK98] Luis A. Caffarelli and Carlos E. Kenig, *Gradient estimates for variable coefficient parabolic equations and singular perturbation problems*, Amer. J. Math. **120** (1998), no. 2, 391–439. MR1613650 (99b:35081)

[CK08] Sagun Chanillo and Carlos E. Kenig, *Weak uniqueness and partial regularity for the composite membrane problem*, J. Eur. Math. Soc. (JEMS) **10** (2008), no. 3, 705–737. MR2421158 (2010b:35496)

[CKS00] Luis A. Caffarelli, Lavi Karp, and Henrik Shahgholian, *Regularity of a free boundary with application to the Pompeiu problem*, Ann. of Math. (2) **151** (2000), no. 1, 269–292. MR1745013 (2001a:35188)

[CKT08] Sagun Chanillo, Carlos E. Kenig, and Tung To, *Regularity of the minimizers in the composite membrane problem in* $\mathbb{R}^2$, J. Funct. Anal. **255** (2008), no. 9, 2299–2320. MR2473259 (2010c:49071)

[CL91] Hi Jun Choe and John L. Lewis, *On the obstacle problem for quasilinear elliptic equations of p Laplacian type*, SIAM J. Math. Anal. **22** (1991), no. 3, 623–638. MR1091673 (92b:35055)

[CPS04] Luis Caffarelli, Arshak Petrosyan, and Henrik Shahgholian, *Regularity of a free boundary in parabolic potential theory*, J. Amer. Math. Soc. **17** (2004), no. 4, 827–869 (electronic). MR2083469 (2005g:35303)

[CR76] L. A. Caffarelli and N. M. Rivière, *Smoothness and analyticity of free boundaries in variational inequalities*, Ann. Scuola Norm. Sup. Pisa Cl. Sci. (4) **3** (1976), no. 2, 289–310. MR0412940 (54 #1061)

[CR77] ______, *Asymptotic behaviour of free boundaries at their singular points*, Ann. of Math. (2) **106** (1977), no. 2, 309–317. MR0463690 (57 #3634)

[Cra84] J. Crank, *Free and moving boundary problems*, Oxford University Press, 1984.

[CRS96] S. J. Chapman, J. Rubinstein, and M. Schatzman, *A mean-field model of superconducting vortices*, European J. Appl. Math. **7** (1996), no. 2, 97–111. MR1388106 (97b:82111)

[CS02] L. Caffarelli and J. Salazar, *Solutions of fully nonlinear elliptic equations with patches of zero gradient: existence, regularity and convexity of level curves*, Trans. Amer. Math. Soc. **354** (2002), no. 8, 3095–3115 (electronic). MR1897393 (2003f:35092)

[CS04] L. A. Caffarelli and H. Shahgholian, *The structure of the singular set of a free boundary in potential theory*, Izv. Nats. Akad. Nauk Armenii Mat. **39** (2004), no. 2, 43–58. MR2167825 (2006e:35347)

[CS05] Luis Caffarelli and Sandro Salsa, *A geometric approach to free boundary problems*, Graduate Studies in Mathematics, vol. 68, American Mathematical Society, Providence, RI, 2005. MR2145284 (2006k:35310)

[CS07] Luis Caffarelli and Luis Silvestre, *An extension problem related to the fractional Laplacian*, Comm. Partial Differential Equations **32** (2007), no. 7-9, 1245–1260. MR2354493 (2009k:35096)

[CSS04] Luis Caffarelli, Jorge Salazar, and Henrik Shahgholian, *Free-boundary regularity for a problem arising in superconductivity*, Arch. Ration. Mech. Anal. **171** (2004), no. 1, 115–128. MR2029533 (2004m:82156)

[CSS08] Luis A. Caffarelli, Sandro Salsa, and Luis Silvestre, *Regularity estimates for the solution and the free boundary of the obstacle problem for the fractional Laplacian*, Invent. Math. **171** (2008), no. 2, 425–461. MR2367025 (2009g:35347)

[Dav74] Philip J. Davis, *The Schwarz function and its applications*, The Mathematical Association of America, Buffalo, N. Y., 1974, The Carus Mathematical Monographs, No. 17. MR0407252 (53 #11031)

[DF86] Emmanuele DiBenedetto and Avner Friedman, *Bubble growth in porous media*, Indiana Univ. Math. J. **35** (1986), no. 3, 573–606. MR855175 (87k:76055)

[DGP07] Donatella Danielli, Nicola Garofalo, and Arshak Petrosyan, *The sub-elliptic obstacle problem: $C^{1,\alpha}$ regularity of the free boundary in Carnot groups of step two*, Adv. Math. **211** (2007), no. 2, 485–516. MR2323535 (2008e:35026)

[DGS03] Donatella Danielli, Nicola Garofalo, and Sandro Salsa, *Variational inequalities with lack of ellipticity. I. Optimal interior regularity and non-degeneracy of the free boundary*, Indiana Univ. Math. J. **52** (2003), no. 2, 361–398. MR1976081 (2004c:35424)

[DL76] G. Duvaut and J.-L. Lions, *Inequalities in mechanics and physics*, Springer-Verlag, Berlin, 1976, Translated from the French by C. W. John, Grundlehren der Mathematischen Wissenschaften, 219. MR0521262 (58 #25191)

[EP08] Anders Edquist and Arshak Petrosyan, *A parabolic almost monotonicity formula*, Math. Ann. **341** (2008), no. 2, 429–454. MR2385663 (2009a:35100)

[ESS98] Charles M. Elliott, Reiner Schätzle, and Barbara E. E. Stoth, *Viscosity solutions of a degenerate parabolic-elliptic system arising in the mean-field theory of superconductivity*, Arch. Ration. Mech. Anal. **145** (1998), no. 2, 99–127. MR1664550 (2000j:35112)

[Eva98] Lawrence C. Evans, *Partial differential equations*, Graduate Studies in Mathematics, vol. 19, American Mathematical Society, Providence, RI, 1998. MR1625845 (99e:35001)

[Eva11] ———, *An introduction to stochastic differential equations*, version 1.2, http://math.berkeley.edu/~evans/SDE.course.pdf, 2011.

[FH76] S. Friedland and W. K. Hayman, *Eigenvalue inequalities for the Dirichlet problem on spheres and the growth of subharmonic functions*, Comment. Math. Helv. **51** (1976), no. 2, 133–161. MR0412442 (54 #568)

[Fic64] Gaetano Fichera, *Problemi elastostatici con vincoli unilaterali: Il problema di Signorini con ambigue condizioni al contorno*, Atti Accad. Naz. Lincei Mem. Cl. Sci. Fis. Mat. Natur. Sez. I (8) **7** (1963/1964), 91–140. MR0178631 (31 #2888)

[Fre72] Jens Frehse, *On the regularity of the solution of a second order variational inequality*, Boll. Un. Mat. Ital. (4) **6** (1972), 312–315. MR0318650 (47 #7197)

[Fre75] ———, *Two dimensional variational problems with thin obstacles*, Math. Z. **143** (1975), no. 3, 279–288. MR0380550 (52 #1450)

[Fre77] ———, *On Signorini's problem and variational problems with thin obstacles*, Ann. Scuola Norm. Sup. Pisa Cl. Sci. (4) **4** (1977), no. 2, 343–362. MR0509085 (58 #22987)

[Fri88] Avner Friedman, *Variational principles and free-boundary problems*, second ed., Robert E. Krieger Publishing Co. Inc., Malabar, FL, 1988. MR1009785 (90k:35271)

[FS86] Avner Friedman and Makoto Sakai, *A characterization of null quadrature domains in $\mathbf{R}^N$*, Indiana Univ. Math. J. **35** (1986), no. 3, 607–610. MR855176 (87k:31005)

[Ger73] Claus Gerhardt, *Regularity of solutions of nonlinear variational inequalities*, Arch. Rational Mech. Anal. **52** (1973), 389–393. MR0377262 (51 #13435)

[GL86] Nicola Garofalo and Fang-Hua Lin, *Monotonicity properties of variational integrals, A_p weights and unique continuation*, Indiana Univ. Math. J. **35** (1986), no. 2, 245–268. MR833393 (88b:35059)

[GL87] ______, *Unique continuation for elliptic operators: a geometric-variational approach*, Comm. Pure Appl. Math. **40** (1987), no. 3, 347–366. MR882069 (88j:35046)

[GP09] Nicola Garofalo and Arshak Petrosyan, *Some new monotonicity formulas and the singular set in the lower dimensional obstacle problem*, Invent. Math. **177** (2009), no. 2, 415–461. MR2511747 (2010m:35574)

[GT01] David Gilbarg and Neil S. Trudinger, *Elliptic partial differential equations of second order*, Classics in Mathematics, Springer-Verlag, Berlin, 2001, Reprint of the 1998 edition. MR1814364 (2001k:35004)

[Gui09] Nestor Guillen, *Optimal regularity for the Signorini problem*, Calc. Var. Partial Differential Equations **36** (2009), no. 4, 533–546. MR2558329 (2010j:35619)

[Gus04] Björn Gustafsson, *Lectures on balayage*, Clifford algebras and potential theory, Univ. Joensuu Dept. Math. Rep. Ser., vol. 7, Univ. Joensuu, Joensuu, 2004, pp. 17–63. MR2103705 (2005k:31001)

[Isa75] V. M. Isakov, *Inverse theorems on the smoothness of potentials*, Differencial′nye Uravnenija **11** (1975), 66–74, 202. MR0361485 (50 #13930)

[Isa76] ______, *Analyticity of the solutions of nonlinear transmission problems*, Differencial′nye Uravnenija **12** (1976), no. 1, 59–68, 187. MR0445123 (56 #3468)

[Isa90] Victor Isakov, *Inverse source problems*, Mathematical Surveys and Monographs, vol. 34, American Mathematical Society, Providence, RI, 1990. MR1071181 (92g:35230)

[Jen80] Robert Jensen, *Boundary regularity for variational inequalities*, Indiana Univ. Math. J. **29** (1980), no. 4, 495–504. MR578201 (81k:35043)

[JK82] David S. Jerison and Carlos E. Kenig, *Boundary behavior of harmonic functions in nontangentially accessible domains*, Adv. in Math. **46** (1982), no. 1, 80–147. MR676988 (84d:31005b)

[Joh48] Fritz John, *Extremum problems with inequalities as subsidiary conditions*, Studies and Essays Presented to R. Courant on his 60th Birthday, January 8, 1948, Interscience Publishers, Inc., New York, N. Y., 1948, pp. 187–204. MR0030135 (10,719b)

[Kar94] Lavi Karp, *On the Newtonian potential of ellipsoids*, Complex Variables Theory Appl. **25** (1994), no. 4, 367–371. MR1314951 (95m:31012)

[Kaw86] Bernhard Kawohl, *On the isoperimetric nature of a rearrangement inequality and its consequences for some variational problems*, Arch. Rational Mech. Anal. **94** (1986), no. 3, 227–243. MR846062 (87i:49018)

[KKPS00] L. Karp, T. Kilpeläinen, A. Petrosyan, and H. Shahgholian, *On the porosity of free boundaries in degenerate variational inequalities*, J. Differential Equations **164** (2000), no. 1, 110–117. MR1761419 (2001c:49017)

[KM96] Lavi Karp and Avmir S. Margulis, *Newtonian potential theory for unbounded sources and applications to free boundary problems*, J. Anal. Math. **70** (1996), 1–63. MR1444257 (98c:35166)

[KM10] ______, *Null quadrature domains and a free boundary problem for the Laplacian*, preprint, 2010.

[KN77] D. Kinderlehrer and L. Nirenberg, *Regularity in free boundary problems*, Ann. Scuola Norm. Sup. Pisa Cl. Sci. (4) **4** (1977), no. 2, 373–391. MR0440187 (55 #13066)

[KN78a] David Kinderlehrer and Louis Nirenberg, *Analyticity at the boundary of solutions of nonlinear second-order parabolic equations*, Comm. Pure Appl. Math. **31** (1978), no. 3, 283–338. MR0460897 (57 #888)

[KN78b] ______, *Hodograph methods and the smoothness of the free boundary in the one phase Stefan problem*, Moving boundary problems (Proc. Sympos. and Workshop, Gatlinburg, Tenn., 1977), Academic Press, New York, 1978, pp. 57–69. MR0481520 (58 #1636)

[KN78c] ______, *The smoothness of the free boundary in the one phase Stefan problem*, Comm. Pure Appl. Math. **31** (1978), no. 3, 257–282. MR480348 (82b:35152)

[KNS78] D. Kinderlehrer, L. Nirenberg, and J. Spruck, *Regularity in elliptic free boundary problems i*, J. Analyse Math. **34** (1978), 86–119 (1979). MR531272 (83d:35060)

[KNS79] ______, *Regularity in elliptic free boundary problems. II. Equations of higher order*, Ann. Scuola Norm. Sup. Pisa Cl. Sci. (4) **6** (1979), no. 4, 637–683. MR563338 (83d:35061)

[KS80] David Kinderlehrer and Guido Stampacchia, *An introduction to variational inequalities and their applications*, Pure and Applied Mathematics, vol. 88, Academic Press Inc. [Harcourt Brace Jovanovich Publishers], New York, 1980. MR567696 (81g:49013)

[Lee98] Ki-ahm Lee, *Obstacle problems for the fully nonlinear elliptic operators*, ProQuest LLC, Ann Arbor, MI, 1998, Thesis (Ph.D.)–New York University. MR2698202

[Lee01] ______, *The obstacle problem for Monge-Ampère equation*, Comm. Partial Differential Equations **26** (2001), no. 1-2, 33–42. MR1842427 (2002d:35069)

[Lew72] Hans Lewy, *On the coincidence set in variational inequalities*, J. Differential Geometry **6** (1972), 497–501, Collection of articles dedicated to S. S. Chern and D. C. Spencer on their sixtieth birthdays. MR0320343 (47 #8882)

[Lio69] J.-L. Lions, *Quelques méthodes de résolution des problèmes aux limites non linéaires*, Dunod, 1969. MR0259693 (41 #4326)

[LS67] J.-L. Lions and G. Stampacchia, *Variational inequalities*, Comm. Pure Appl. Math. **20** (1967), 493–519. MR0216344 (35 #7178)

[LS69] Hans Lewy and Guido Stampacchia, *On the regularity of the solution of a variational inequality*, Comm. Pure Appl. Math. **22** (1969), 153–188. MR0247551 (40 #816)

[LS70] ______, *On the smoothness of superharmonics which solve a minimum problem*, J. Analyse Math. **23** (1970), 227–236. MR0271383 (42 #6266)

[LS71] ______, *On existence and smoothness of solutions of some non-coercive variational inequalities*, Arch. Rational Mech. Anal. **41** (1971), 241–253. MR0346313 (49 #11038)

[LS03] K. Lee and H. Shahgholian, *Hausdorff measure and stability for the p-obstacle problem* ($2 < p < \infty$), J. Differential Equations **195** (2003), no. 1, 14–24. MR2019240 (2004k:35423)

[Mar82] A. S. Margulis, *On potential theory in the classes* $L_p(\Omega)$, Izv. Vyssh. Uchebn. Zaved. Mat. (1982), no. 1, 33–41. MR651659 (83g:31005)

[Mat05] Norayr Matevosyan, *Tangential touch between free and fixed boundaries in a problem from superconductivity*, Comm. Partial Differential Equations **30** (2005), no. 7-9, 1205–1216. MR2180300 (2006f:35305)

[Mon03] R. Monneau, *On the number of singularities for the obstacle problem in two dimensions*, J. Geom. Anal. **13** (2003), no. 2, 359–389. MR1967031 (2004a:35223)

[Mon04] Régis Monneau, *On the regularity of a free boundary for a nonlinear obstacle problem arising in superconductor modelling*, Ann. Fac. Sci. Toulouse Math. (6) **13** (2004), no. 2, 289–311. MR2126745 (2006a:35320)

[Mon09] R. Monneau, *Pointwise estimates for Laplace equation. Applications to the free boundary of the obstacle problem with Dini coefficients*, J. Fourier Anal. Appl. **15** (2009), no. 3, 279–335. MR2511866 (2010h:35064)

[Mor08] Charles B. Morrey, Jr., *Multiple integrals in the calculus of variations*, Classics in Mathematics, Springer-Verlag, Berlin, 2008, Reprint of the 1966 edition [MR0202511]. MR2492985

[MP11] Norayr Matevosyan and Arshak Petrosyan, *Almost monotonicity formulas for elliptic and parabolic operators with variable coefficients*, Comm. Pure Appl. Math. **64** (2011), no. 2, 271–311. MR2766528

[MW07] R. Monneau and G. S. Weiss, *An unstable elliptic free boundary problem arising in solid combustion*, Duke Math. J. **136** (2007), no. 2, 321–341. MR2286633 (2007k:35527)

[Ou94] Biao Ou, *Global solutions to a free boundary problem*, Comm. Partial Differential Equations **19** (1994), no. 3-4, 369–397. MR1265804 (95e:35233)

[Pom29] D. Pompéiu, *Sur certains systèmes d'équations linéaires et sur une propriété intégrale des fonctions de plusieurs variables*, C. R. **188** (1929), 1138–1139.

[PS07] Arshak Petrosyan and Henrik Shahgholian, *Geometric and energetic criteria for the free boundary regularity in an obstacle-type problem*, Amer. J. Math. **129** (2007), no. 6, 1659–1688. MR2369892 (2008m:35384)

[Ric78] David Joseph Allyn Richardson, *Variational problems with thin obstacles*, ProQuest LLC, Ann Arbor, MI, 1978, Thesis (Ph.D.)–The University of British Columbia (Canada). MR2628343

[Rod87] José-Francisco Rodrigues, *Obstacle problems in mathematical physics*, North-Holland Mathematics Studies, vol. 134, North-Holland Publishing Co., Amsterdam, 1987, Notas de Matemática [Mathematical Notes], 114. MR880369 (88d:35006)

[Sak81] Makoto Sakai, *Null quadrature domains*, J. Analyse Math. **40** (1981), 144–154 (1982). MR659788 (84e:30069)

[Sak82] ———, *Quadrature domains*, Lecture Notes in Mathematics, vol. 934, Springer-Verlag, Berlin, 1982. MR663007 (84h:41047)

[Sak91] ———, *Regularity of a boundary having a Schwarz function*, Acta Math. **166** (1991), no. 3-4, 263–297. MR1097025 (92c:30042)

[Sak93] ———, *Regularity of boundaries of quadrature domains in two dimensions*, SIAM J. Math. Anal. **24** (1993), no. 2, 341–364. MR1205531 (94c:30054)

[Sar75] Jukka Sarvas, *The Hausdorff dimension of the branch set of a quasiregular mapping*, Ann. Acad. Sci. Fenn. Ser. A I Math. **1** (1975), no. 2, 297–307. MR0396945 (53 #805)

[Sch77] David G. Schaeffer, *Some examples of singularities in a free boundary*, Ann. Scuola Norm. Sup. Pisa Cl. Sci. (4) **4** (1977), no. 1, 133–144. MR0516201 (58 #24345)

[Sch89] Rainer Schumann, *Regularity for Signorini's problem in linear elasticity*, Manuscripta Math. **63** (1989), no. 3, 255–291. MR986184 (90c:35098)

[Sha92] Henrik Shahgholian, *On quadrature domains and the Schwarz potential*, J. Math. Anal. Appl. **171** (1992), no. 1, 61–78. MR1192493 (93m:31006)

[Sha03] ______, *$C^{1,1}$ regularity in semilinear elliptic problems*, Comm. Pure Appl. Math. **56** (2003), no. 2, 278–281. MR1934623 (2003h:35087)

[Sha07] ______, *The singular set for the composite membrane problem*, Comm. Math. Phys. **271** (2007), no. 1, 93–101. MR2283955 (2008e:35214)

[Sig59] A. Signorini, *Questioni di elasticità non linearizzata e semilinearizzata*, Rend. Mat. e Appl. (5) **18** (1959), 95–139. MR0118021 (22 #8794)

[Sil07] Luis Silvestre, *Regularity of the obstacle problem for a fractional power of the Laplace operator*, Comm. Pure Appl. Math. **60** (2007), no. 1, 67–112. MR2270163 (2008a:35041)

[Spr83] Joel Spruck, *Uniqueness in a diffusion model of population biology*, Comm. Partial Differential Equations **8** (1983), no. 15, 1605–1620. MR729195 (85h:35086)

[SS00] Étienne Sandier and Sylvia Serfaty, *A rigorous derivation of a free-boundary problem arising in superconductivity*, Ann. Sci. École Norm. Sup. (4) **33** (2000), no. 4, 561–592. MR1832824 (2002k:35324)

[Sta64] Guido Stampacchia, *Formes bilinéaires coercitives sur les ensembles convexes*, C. R. Acad. Sci. Paris **258** (1964), 4413–4416. MR0166591 (29 #3864)

[Str74] V. N. Strakhov, *The inverse logarithmic potential problem for contact surface*, Izv. Acad. Sci. USSR Phys. Solid Earth **10** (1974), 104–114 (Translated from the Russian).

[SU03] Henrik Shahgholian and Nina Uraltseva, *Regularity properties of a free boundary near contact points with the fixed boundary*, Duke Math. J. **116** (2003), no. 1, 1–34. MR1950478 (2003m:35253)

[SUW04] Henrik Shahgholian, Nina Uraltseva, and Georg S. Weiss, *Global solutions of an obstacle-problem-like equation with two phases*, Monatsh. Math. **142** (2004), no. 1-2, 27–34. MR2065019 (2005c:35301)

[SUW07] ______, *The two-phase membrane problem—regularity of the free boundaries in higher dimensions*, Int. Math. Res. Not. IMRN (2007), no. 8, Art. ID rnm026, 16. MR2340105 (2009b:35444)

[SUW09] ______, *A parabolic two-phase obstacle-like equation*, Adv. Math. **221** (2009), no. 3, 861–881. MR2511041 (2010f:35441)

[SW06] Henrik Shahgholian and Georg S. Weiss, *The two-phase membrane problem—an intersection-comparison approach to the regularity at branch points*, Adv. Math. **205** (2006), no. 2, 487–503. MR2258264 (2007k:35167)

[Ura85] N. N. Uraltseva, *Hölder continuity of gradients of solutions of parabolic equations with boundary conditions of Signorini type*, Dokl. Akad. Nauk SSSR **280** (1985), no. 3, 563–565. MR775926 (87b:35025)

[Ura96] ______, *C^1 regularity of the boundary of a noncoincident set in a problem with an obstacle*, St. Petersburg Math. J. **8** (1997), no. 2, 341–353. MR1392033 (97m:35105)

[Ura01] ______, *Two-phase obstacle problem*, J. Math. Sci. (New York) **106** (2001), no. 3, 3073–3077, Function theory and phase transitions. MR1906034 (2003e:35331)

[Ura07] ______, *Boundary estimates for solutions of elliptic and parabolic equations with discontinuous nonlinearities*, Nonlinear equations and spectral theory, Amer. Math. Soc. Transl. Ser. 2, vol. 220, Amer. Math. Soc., Providence, RI, 2007, pp. 235–246. MR2343613 (2008k:35040)

[Wei99a] G. S. Weiss, *Self-similar blow-up and Hausdorff dimension estimates for a class of parabolic free boundary problems*, SIAM J. Math. Anal. **30** (1999), no. 3, 623–644 (electronic). MR1677947 (2000d:35267)

[Wei99b] Georg S. Weiss, *A homogeneity improvement approach to the obstacle problem*, Invent. Math. **138** (1999), no. 1, 23–50. MR1714335 (2000h:35057)

[Wei01] G. S. Weiss, *An obstacle-problem-like equation with two phases: pointwise regularity of the solution and an estimate of the Hausdorff dimension of the free boundary*, Interfaces Free Bound. **3** (2001), no. 2, 121–128. MR1825655 (2002c:35275)

[WHD95] Paul Wilmott, Sam Howison, and Jeff Dewynne, *The mathematics of financial derivatives*, Cambridge University Press, Cambridge, 1995, A student introduction. MR1357666 (96h:90028)

[Whi34] Hassler Whitney, *Analytic extensions of differentiable functions defined in closed sets*, Trans. Amer. Math. Soc. **36** (1934), no. 1, 63–89. MR1501735

[Wid67] Kjell-Ove Widman, *Inequalities for the Green function and boundary continuity of the gradient of solutions of elliptic differential equations*, Math. Scand. **21** (1967), 17–37 (1968). MR0239264 (39 #621)

[Wil76] Stephen A. Williams, *A partial solution of the Pompeiu problem*, Math. Ann. **223** (1976), no. 2, 183–190. MR0414904 (54 #2996)

[Wil81] ———, *Analyticity of the boundary for Lipschitz domains without the Pompeiu property*, Indiana Univ. Math. J. **30** (1981), no. 3, 357–369. MR611225 (82j:31009)

[Yau82] Shing Tung Yau, *Problem section*, Seminar on Differential Geometry, Ann. of Math. Stud., vol. 102, Princeton Univ. Press, Princeton, N.J., 1982, pp. 669–706. MR645762 (83e:53029)

Notation

Basic notation

$c, c_0, C, C_0, C_1, \ldots$	generic constants
$C_{a_1,\ldots,a_k}, C(a_1,\ldots,a_k)$	constants depending (only) on $a_1, \ldots, a_k$
$\mathbb{N}$	$\{1,2,3,\ldots\}$, the set of natural numbers
$\mathbb{R}$	$(-\infty,\infty)$, the set of real numbers
$\mathbb{R}^n$	$\{x=(x_1,\ldots,x_n) : x_i \in \mathbb{R}, i=1,\ldots,n\}$, $n \in \mathbb{N}$, the n-dimensional Euclidean space
$e_1, e_2, \ldots, e_n$	$e_i = (0,\ldots,1,\ldots,0)$, with 1 in the i-th position, the standard coordinate vectors in $\mathbb{R}^n$
$x \cdot y$	$\sum_{i=1}^n x_i y_i$, the interior product of $x, y \in \mathbb{R}^n$
$\lvert x \rvert$	$\sqrt{x \cdot x}$ for $x \in \mathbb{R}^n$, the Euclidean norm
x'	$(x_1,\ldots,x_{n-1})$ for $x = (x_1,\ldots,x_n) \in \mathbb{R}^n$; we also identify $x = (x', x_n)$
$\mathbb{R}^n_+, \mathbb{R}^n_-$	$\{x \in \mathbb{R}^n : x_n > 0\}$; $\{x \in \mathbb{R}^n : x_n < 0\}$
$B_r(x)$	$\{y \in \mathbb{R}^n : \lvert y-x \rvert < r\}$, the open ball in $\mathbb{R}^n$
$B_r^\pm(x)$	$B_r(x) \cap \mathbb{R}^n_\pm$
$B_r, B_r^\pm$	$B_r(0)$, $B_r^\pm(0)$
B'_r	$\{y' \in \mathbb{R}^{n-1} : \lvert y' \rvert < r\}$, ball in $\mathbb{R}^{n-1}$; often identified with $B_r \cap (\mathbb{R}^{n-1} \times \{0\})$
$\mathcal{C}_\delta$	$\{x \in \mathbb{R}^n : x_n > \delta \lvert x' \rvert\}$, for $\delta > 0$, cone with axis e_n and opening angle $2\arctan(1/\delta)$
∂E, $\overline{E}$, $\mathrm{Int}(E)$	the boundary, closure, and the interior of the set E in the relevant topology

E^c	$\mathbb{R}^n \setminus E$, the complement of the set $E \subset \mathbb{R}^n$
$\operatorname{dist}(x, E)$	$\inf_{y\in E} \lvert x-y\rvert$ for $x \in \mathbb{R}^n$, $E \subset \mathbb{R}^n$
$\operatorname{diam}(E)$	$\sup_{x,y\in E} \lvert x-y\rvert$ for $E \subset \mathbb{R}^n$
$\lvert E\rvert$	Lebesgue measure of a measurable set $E \subset \mathbb{R}^n$
$H^s(E)$, $s \leq 0 \leq n$	s-Hausdorff measure for a Borel set E
χ_E	$\begin{cases} 1 & \text{on } E, \\ 0 & \text{on } E^c, \end{cases}$ the characteristic function of $E \subset \mathbb{R}^n$
$A \Subset B$	$\overline{A}$ is compact and contained in B
a^+, a^-	$\max\{a, 0\}$; $\max\{-a, 0\}$ for a number $a \in \mathbb{R}$; for a function $u : E \to \mathbb{R}$, $u^\pm : E \to \mathbb{R}_+$ are given by $u^\pm(x) = u(x)^\pm$
$\operatorname{supp} u$	$\overline{\{u \neq 0\}}$, the support of a function u
$\partial_e u$, u_e	the partial derivative of a function u in direction e, understood in the classical sense or in the sense of distributions
$\partial_{x_i} u$, u_{x_i}	same as $\partial_{e_i} u$, $i = 1, 2, \ldots, n$, where e_i are the standard coordinate vectors
$\partial_{x_{i_1}\cdots x_{i_k}} u$, $u_{x_{i_1}\cdots x_{i_k}}$	$\partial_{x_{i_1}} \partial_{x_{i_2}} \cdots \partial_{x_{i_k}} u$, higher order derivatives
$\partial^\beta u$	$\partial^{\beta_1}_{x_1} \cdots \partial^{\beta_k}_{x_k} u$, for multiindex $\beta = (\beta_1, \ldots, \beta_n)$, $\beta_i \in \mathbb{N} \cup \{0\}$, partial derivative of order $\lvert\beta\rvert = \beta_1 + \cdots + \beta_n$
Δu	$\sum_{i=1}^n \partial_{x_i x_i} u$, the Laplace operator (Laplacian) in $\mathbb{R}^n$
∇u	$(\partial_{x_1} u, \ldots, \partial_{x_n} u)$, the gradient of u
$D^2 u$	$(\partial_{x_i x_j} u)_{i,j=1}^n$, the Hessian matrix of u
$D^k u$	for $k \in \mathbb{N}$, $(\partial_{x_{i_1}\cdots x_{i_k}} u)$, the tensor of partial derivatives of k-th order

Function spaces

$C^0(E)$: the Banach space of continuous function $u : E \to \mathbb{R}$ on the set E with the uniform norm

$$\|u\|_{C^0(E)} = \sup_{x\in E} |u(x)|.$$

$C(E)$: same as $C^0(E)$.

$C^{0,\alpha}(E)$, $\alpha \in (0, 1]$: the Banach space of α-Hölder continuous functions of E with the norm

$$\|u\|_{C^{0,\alpha}(E)} = \|u\|_{C^0(E)} + [u]_{\alpha,E}, \quad [u]_{\alpha,E} = \sup \frac{|u(x) - u(y)|}{|x-y|^\alpha}.$$

The quantity $[u]_{\alpha,E}$ is called the α-Hölder seminorm of u on E. When $\alpha = 1$, $C^{0,1}$ is the space of Lipschitz continuous functions and $[u]_{1,E}$ is called the Lipschitz constant of u on E.

$C^\alpha(E)$, $\alpha \in (0, 1)$: same as $C^{0,\alpha}(E)$; we never use this notation for $\alpha = 1$.

$C^k(D)$, $k \in \mathbb{N}$: for open $D \subset \mathbb{R}^n$, the Fréchet space of functions $u : D \to \mathbb{R}$ with continuous partial derivatives $\partial^\beta u$ in D, $|\beta| \leq k$; it has the family of seminorms

$$\|u\|_{C^k(K)} = \sum_{|\beta| \leq k} \|\partial^\beta u\|_{C^0(K)} \quad \text{for } K \Subset D.$$

$C^\infty(D)$: the Fréchet space of infinitely differentiable functions in D, intersection of all $C^k(D)$, $k \in \mathbb{N}$.

$C_0^k(D)$, $k \in \mathbb{N} \cup \{0, \infty\}$: a subspace of $C^k(D)$ of functions compactly supported in D, i.e. with $\operatorname{supp} u \Subset D$.

$C^{k,\alpha}(D)$, $k \in \mathbb{N}$, $\alpha \in (0,1]$: the Fréchet space consisting of functions $u \in C^k(D)$ such that $\partial^\beta u \in C^{0,\alpha}(K)$, $|\beta| = k$, for any $K \Subset D$, with the family of seminorms

$$\|u\|_{C^{k,\alpha}(K)} = \|u\|_{C^k(K)} + \sum_{|\beta|=k} [\partial^\beta u]_{\alpha,K}.$$

$C^k(D \cup F)$, $C^{k,\alpha}(D \cup F)$: for a relatively open $F \subset \partial D$, similar to $C^k(D)$ and $C^{k,\alpha}(D)$ but with the k-th partial derivatives continuous (α-Hölder continuous) up to F and with the respective seminorms for $K \Subset D \cup F$.

$C^{k,\alpha}_{\mathrm{loc}}(D)$, $C^{k,\alpha}_{\mathrm{loc}}(D \cup F)$: same as $C^{k,\alpha}(D)$, $C^{k,\alpha}(D \cup F)$; we use this notation in the instances where we want to emphasise the local nature of the estimates.

$L^p(E)$, $1 \leq p < \infty$: for a measurable set $E \subset \mathbb{R}^n$, the Banach space of measurable functions u on E (more precisely, classes of a.e. equivalences) with the finite norm

$$\|u\|_{L^p(E)} = \left(\int_E |u(x)|^p dx \right)^{1/p}.$$

$L^\infty(E)$: the Banach space of essentially bounded measurable functions u with the norm

$$\|u\|_{L^\infty(E)} = \operatorname*{ess\,sup}_{x \in E} |u(x)|.$$

$W^{k,p}(D)$, $k \in \mathbb{N}$, $1 \leq p \leq \infty$: the Sobolev space, the Banach space of functions $u \in L^p(D)$ with the (distributional) partial derivative $\partial^\beta u \in L^p(D)$, $|\beta| \leq k$; the norm is given by

$$\|u\|_{W^{k,p}(D)} = \sum_{|\beta| \leq k} \|\partial^\beta u\|_{L^p(D)}.$$

$W_0^{k,p}(D)$, $k \in \mathbb{N}$, $1 \leq p \leq \infty$: the closure of $C_0^\infty(D)$ in $W^{k,p}(D)$.

$L^p_{\mathrm{loc}}(D)$, $W^{k,p}_{\mathrm{loc}}(D)$, $k \in \mathbb{N}$, $1 \leq p \leq \infty$: the Fréchet space of measurable functions $u : D \to \mathbb{R}$ such that $u|_{D'} \in L^p(D')$ or, respectively, $W^{k,p}(D')$ for any open $D' \Subset D$.

Notation related to free boundaries

$\delta(\rho, u, x^0)$ — the thickness function for the coincidence set for the solution u at x^0, defined as $\frac{1}{\rho}\operatorname{min\,diam}(\Omega(u)^c \cap B_\rho(x^0))$

$\delta(\rho, u)$ — same as $\delta(\rho, u, 0)$

$\Gamma_\pm(u)$ — in Problem **C**, the free boundaries of positive and negative phases, $\partial\Omega^\pm(u) \cap D$

$\Gamma^0(u)$ — the set of branch points in Problem **C**, defined as $\Gamma(u) \cap \{|\nabla u| = 0\}$

$\Gamma^*(u)$ — the set of nonbranch points in Problem **C**, defined as $\Gamma(u) \cap \{|\nabla u| \neq 0\}$

$\Gamma_\kappa(u)$ — the set of free boundary points in Problem **S**, where the blowups have homogeneity κ, i.e. $\{x^0 \in \Gamma(u) : N(0+, u, x^0) = \kappa\}$

$\Lambda(u)$, Λ — the coincidence set; in the classical obstacle problem, the set $\{x \in D : u(x) = \psi(x)\}$; in Problem **A**, the set $\{x \in D : u(x) = |\nabla u(x)| = 0\}$; in Problem **B**, the set $\{x \in D : |\nabla u(x)| = 0\}$; in the thin obstacle problem, the set $\{x \in \mathcal{M} : u(x) = \psi(x)\}$; in Problem **S**, the set $\{x' \in B'_R : u(x', 0) = 0\}$

$\operatorname{min\,diam} E$ — the minimum diameter of the set E, the infimum of distances between pairs of parallel planes enclosing E

$M(r, u, p)$, $M(r, u, p, x^0)$ — Monneau's monotonicity formula for the classical obstacle problem; see p. 138

$M_\kappa(r, u, p)$, $M_\kappa(r, u, p, x^0)$ — Monneau-type monotonicity formula for Problem **S**; see p. 189

$N(r, u)$, $N(r, u, x^0)$ — Almgren's frequency function (at x^0); see pp. 173, 178

$\Omega(u)$, Ω — the complement of the coincidence set $\Lambda(u)$

$\Omega_\pm(u)$ — in Problem **C**, regions of positive and negative phase defined as $\{x \in D : \pm u(x) > 0\}$

$P_R(x^0, M)$ — a class of local solutions of Problems **A**, **B**, **C** in $B_R(x^0)$; see p. 65

$P_R(M)$ — same as $P_R(0, M)$

$P_R^+(x^0, M)$, $P_R^+(M)$ — classes of local solutions of Problems **A**, **B**, **C** in B_R^+ (see p. 154); $P_R^+(M)$ should not be confused with $P_R^+(0, M)$

$P_\infty(M)$	a class of global solutions of Problems **A**, **B**, **C**; see p. 65
$P_\infty^+(M)$	a class of solutions of Problems **A**, **B**, **C** in $\mathbb{R}^n_+$; see p. 154
$\Sigma(u)$	the set of singular points; for Problem **A**, see p. 134; for Problem **S**, see p. 187
$\Sigma_\kappa(u)$	for $\kappa = 2m$, $m \in \mathbb{N}$, the same as $\Gamma_\kappa(u)$
$\Sigma^d(u)$, $\Sigma^d_\kappa(u)$	the set of singular points where the dimension of the singular set equals d; for classical obstacle problem, see p. 141; for Problem **A**, see p. 146; for Problem **S**, see p. 192
$u_{x^0,\lambda}(x)$	a rescaling of solution u; in Problems **A**, **B**, **C** defined as $u_{x^0,\lambda}(x) := \frac{1}{\lambda^2}(u(x^0+\lambda x)$; in Problem **S** defined as $u_{x^0,\lambda}(x) := \frac{u(x^0+\lambda x)}{\left(\lambda^{-(n-1)}\int_{\partial B_\lambda(x^0)} u^2\right)^{1/2}}$
$u_\lambda(x)$	same as $u_{x^0,\lambda}(x)$ when the scaling center x^0 is unambiguously identified from the context
$u_\lambda^{(\kappa)}(x)$, $u_{x^0,\lambda}^{(\kappa)}(x)$	κ-homogeneous rescalings in Problem **S**, defined as $\frac{1}{\lambda^\kappa}u(x^0+\lambda x)$
$W(r,u)$, $W(r,u,x^0)$	Weiss's energy functional (at x^0) in Problems **A**, **B**, **C**; see p. 74
$W_\kappa(r,u)$, $W_\kappa(r,u,x^0)$	Weiss's energy functional (at x^0) in Problem **S**; see p. 188
$\Phi(r,u_+,u_-)$	the Alt-Caffarelli-Friedman functional; see p. 34
$\phi_e(r,u)$	$\Phi(r,(\partial_e u)^+,(\partial_e u)^-)$

Index